Lecture Notes in Statistics 194

Edited by P. Bickel, P.J. Diggle, S. Fienberg,
U. Gather, I. Olkin, S. Zeger

For other titles published in this series, go to
www.springer.com/series/694

Dario Basso · Fortunato Pesarin · Luigi Salmaso ·
Aldo Solari

Permutation Tests
for Stochastic Ordering
and ANOVA

Theory and Applications with R

 Springer

Dario Basso
Università di Padova
Dip. to Tecnica e Gestione dei
Sistemi Industriali
Stradella San Nicola, 3
36100 Vicenza
Italy
basso@gest.unipd.it

Fortunato Pesarin
Università di Padova
Dip. to Scienze Statistiche
Via Cesare Battisti, 241/243
35121 Padova
Italy
pesarin@stat.unipd.it

Luigi Salmaso
Università di Padova
Dip. to Tecnica e Gestione dei
Sistemi Industriali
Stradella San Nicola, 3
36100 Vicenza
Italy
salmaso@gest.unipd.it

Aldo Solari
Università di Padova
Dip. to Processi Chimici
dell' Ingegneria
via Marzolo, 9
35131 Padova
Italy
aldo.solari@unipd.it

ISBN 978-0-387-85955-2   e-ISBN 978-0-387-85956-9
DOI 10.1007/978-0-387-85956-9
Springer Dordrecht Heidelberg London New York

Library of Congress Control Number: PCN applied for

Printed on acid-free paper

Springer is part of Springer Science+Business Media (www.springer.com)

Preface

This book is divided into two parts. The first part is devoted to some advances in testing for a stochastic ordering, and the second part is related to ANOVA procedures for nonparametric inference in experimental designs. It is worth noting that, before introducing specific arguments in the two main parts of the book, we provide an introductory first chapter on basic theory of univariate and multivariate permutation tests, with a special look at multiple-comparison and multiple testing procedures.

The concept of stochastic ordering of distributions was introduced by Lehmann (1955) and plays an important role in the theory of statistical inference. It arises in many applications in which it is believed that, given a response variable Y and an explanatory variable x, the statistical model assumes that the distribution of $Y|x$ belongs to a certain family of probability distributions that is ordered in the sense, roughly speaking, that large values of x lead to large values of the Y's.

Many types of orderings of varying degrees of strength have been defined in the literature to compare the order of magnitude of two or more distributions (see Shaked and Shanthikumar, 1994, for a review). These include likelihood ratio ordering, hazard rate ordering, and simple stochastic ordering, which are perhaps the main instances. On the one hand, these orderings make the statistical inference procedures more complicated. On the other, they contain statistical information as well, so that if properly incorporated they would be more efficient than their counterparts, wherein such constraints are ignored. These considerations emphasize the importance of statistical procedures to detect the occurrence of such orderings on the basis of random samples. Inference based on stochastic orderings for univariate distributions has been studied extensively, whereas for multivariate distributions it has received much less attention because the "curse of dimensionality" makes the statistical procedures considerably more complicated. For a review of constrained inference, we refer to the recent monograph by Silvapulle and Sen (2005).

Likelihood inference is perhaps the default methodology for many statistical problems; indeed, the overwhelming majority of work related to order-restricted problems is based on the likelihood principle. However, there are instances when one might prefer a competitive procedure. Recently there have been debates about the suitability of different test procedures: Perlman and Chaudhuri (2004a) argue in favor of likelihood ratio tests, whereas Cohen and Sackrowitz (2004) argue in favor of the so-called class of directed tests. In multidimensional problems, it is rare that a "best" inference procedure exists. However, even in such a complex setup, following Roy's union-intersection principle (Roy, 1953), it might be possible to look upon the null hypothesis as the intersection of several component hypotheses and the alternative hypothesis as the union of the same number of component alternatives, giving rise to a multiple testing problem. A classical approach is to require that the probability of rejecting one or more true null hypotheses, the familywise error rate (Hochberg and Tamhane, 1987), not exceed a given level. Generally, it is surprising that some existing procedures seem to be satisfied to stop with a global test just dealing with the acceptance or rejection of the intersection of all null hypotheses. In the form presented, it will be difficult to interpret a statistically significant finding: The statistical significance of the individual hypotheses in multiple-endpoint or multiple-comparison problems remains very important even if global tests indicate an overall effect. Indeed, most clinical trials are conducted to compare a treatment group with a control group on multiple endpoints, and the inferential goal after establishing an overall treatment effect is to identify the individual endpoints on which the treatment is better than the control. For tests of equality of means in a one-way classification, the ANOVA F test is available, but in the case of rejection of the global null hypothesis of equality of all means, one will frequently want to know more about the means than just that they are unequal.

In the majority of the situations we shall deal with, both the hypothesis and the class of alternatives may be nonparametric, and as a result it may be difficult even to construct tests that satisfactorily control the level (exactly or asymptotically). For such situations, we will consider permutation methods that achieve this goal under fairly general assumptions. Under exchangeability of the data, the empirical distribution of the values of a given statistic recomputed over transformations of the data serves as a null distribution; this leads to exact control of the level in such models. In addition, by making effective use of resampling to implicitly estimate the dependence structure of multiple test statistics, it is possible to construct valid and efficient multiple testing procedures that strongly control the familywise error rate, as in Westfall and Young (1993).

We bring out the permutation approach for models in which there is a possibly multivariate response vector $\boldsymbol{Y}$ and an ordinal explanatory variable x taking values $\{1, \ldots, k\}$, which can be thought of as several levels of a treatment. Let $\boldsymbol{Y}_i$ denote the random vector whose distribution is the conditional distribution of $\boldsymbol{Y}$ given $x = i$. We are interested in testing $\boldsymbol{Y}_1 \overset{\mathrm{d}}{=} \ldots \overset{\mathrm{d}}{=} \boldsymbol{Y}_k$ against a stochastic ordering alternative $\boldsymbol{Y}_1 \overset{\mathrm{st}}{\leq} \ldots \overset{\mathrm{st}}{\leq} \boldsymbol{Y}_k$ with at least one $\overset{\mathrm{st}}{\lneq}$.

In the statistical literature, there is relatively little on multivariate models for nonnormal response variables, such as ordinal response data. This is perhaps due to the mathematical intractability of reasonable models and to related computational problems. The aim is therefore to provide permutation methods that apply to multivariate discrete and continuous data. We deal with univariate and multivariate ordinal data in Chapters 2 and 3, respectively, and Chapter 4 contains results for multivariate continuous responses.

As previously said, the second part of the book is dedicated to nonparametric ANOVA within the permutation framework. Experimental designs are useful research tools that are applied in almost all scientific fields. In factorial experiments, processes of various natures whose behavior depends on several factors are studied. In this context, a factor is any characteristic of the experimental condition that might influence the results of the experiment. Every factor takes on different values, called levels, that can be either quantitative (dose) or qualitative (category). When several factors are observed in an experiment, every possible combination of their levels is called a treatment. The analysis of factorial designs through linear models allows us to study (and assess) the effect of the experimental factors on the response, where factors are under the control of the experimenter. They also allow for evaluating the joint effect of two or more factors (also named main factors), which are known as interaction factors. The statistical analysis is usually carried on by assuming a linear model to fit the data. Here, the model to fit the response is an additive model, where the effect of main factors and interactions are represented by unknown parameters. In addition, a stochastic error component is considered in order to represent the inner variability of the response. Usually, errors are assumed to be i.i.d. homoscedastic random variables with zero mean. This model requires some further assumptions in order to be applied. Some of them, such as independence among experimental units or the identical distribution, are reasonable and supported by experience. Other assumptions, such as normality of the experimental errors, are not always adequate. Generally it is possible to check the assumption of normality only after the analysis has been made, through diagnostic tools such as the $Q - Q$ plot (Daniel, 1959). Nevertheless, these tools are mainly descriptive; therefore the conclusions they may lead to are essentially subjective. If the normality of errors is not satisfied or cannot be justified, then the usual test statistics (such as the Student t test or the F test) are approximate. It is therefore worthwhile to reduce some assumptions, either to avoid the use of approximate tests or to extend the applicability of the methods applied.

Permutation tests represent the ideal instrument in the experimental design field since they do not require assumptions on the distribution of errors and, if normality can be assumed, they give results almost as powerful as their parametric counterpart. There are other reasons to use permutation tests; for instance, in the $I \times J$ replicated designs, even if data are normally distributed, the two-way ANOVA test statistics are positively correlated. This means that the inference on one factor may be influenced by other factors. There are other situations where parametric tests cannot be applied at all: In unreplicated full factorial designs, the number of observations equals the number of parameters to estimate in the model; therefore there are no degrees of freedom left to estimate the error variance. Permutation tests deal with the notion of exchangeability of the responses: The exchangeability is satisfied if the probability of the observed data is invariant with respect to random permutations of the indexes. The exchangeability of the responses is a sufficient condition to obtain an exact inference. In factorial design, the responses are generally not exchangeable since units assigned to different treatments have different expectations. Thus, either a restricted kind of permutation is needed or approximate solutions must be taken into account in order to obtain separate inferences on the main factor/interaction effects.

Chapter 5 is an introduction to ANOVA in a nonparametric view. Therefore, the general layout is introduced with minimal assumptions, with some particular care about the exchangeability of errors. Some of the solutions from the literature are introduced and discussed. The kinds of errors that may arise (individual and family wise errors) in such a context are introduced, and some preliminary methods to control them are suggested. The final part of the chapter leads with direct applications of the existing methods from the literature to practical examples.

In Chapter 6 a nonparametric solution to test for effects in replicated designs is introduced. This part is dedicated to extending the solution proposed by Pesarin (2001) and Salmaso (2003) for a 2×2 balanced replicated factorial design with n units per treatment. Since the responses are not exchangeable, the solution is based on a particular kind of permutations, named synchronized permutations. In particular, by exchanging units within the same level of a factor and by assuming the standard side conditions on the constraints, it is possible to obtain a test statistic for main factors and interactions that only depends on the effects under testing and on a combination of exchangeable errors. The proposed tests are uncorrelated with each other, and they are shown to be almost as powerful as the two-way ANOVA test statistics when errors are normally distributed. After introducing the test statistics, two algorithms are proposed to obtain Monte Carlo synchronized permutations. If we desire a post hoc comparison, simultaneous confidence intervals on all pair wise comparisons can be obtained by similarly applying synchronized permutations. The tests proposed are then compared with the classical parametric analysis.

Chapter 7 is devoted to the problem of the unreplicated full factorial design analysis. Again, the problem of exchangeability of the responses arises and, given the peculiarity of the problem, it does not seem possible to obtain exact permutation tests for all factors unless testing for the global null hypothesis that there are no treatment effects. The paired permutation test introduced by Pesarin and Salmaso (2002) is exact, but it is only applicable to the first M largest effects. A further approximate solution is then proposed. Such a solution is based on the decomposition of the total response variance under the full model and under some restricted models that are obtained in accordance with the null hypothesis under testing. The test statistic is a ratio of uncorrelated random variables, that allows us to evaluate the increase of explained variance in the full model due to the main effect under testing. The proposed test statistic allows the individual error rate to be controlled under the effect sparsity assumption. It does not control the experimental error rate, and its power is a decreasing function of the number of active effects and their sizes (the bigger the size of one effect, the bigger the noncentrality parameter in the denominator of the test statistic). To allow of control the experiment-wise error rate and in order to gain power, another version of the statistical procedure is introduced, a step-up procedure based on the comparison among noncentrality parameters of the estimates of factor effects. This test needs a calibration, which requires the central limit theorem, in order to control the experiment-wise error rate. The calibration can be obtained by either providing some critical p-values for each step of the procedure in accordance with a Bonferroni (or Bonferroni-Holm) correction or by obtaining a single critical p-value based on the distribution of the minP from simulated data under the global null hypothesis. This test is shown to be very powerful, as it can detect active factors even when there is no effect sparsity assumption (except on the smallest estimated effect, which cannot be tested). Note that a similar calibration can be provided in order to control the individual error rate at level α by choosing the critical α-quantile from the simulated null distribution of the sequential p-values. A power comparison with Loughin and Noble's test (1997) and an application from Montgomery (1991) are finally reported and discussed.

Each chapter of the book contains R code to develop the proposed theory. All R codes and related functions are available online at www.gest.unipd.it/~salmaso/web/springerbook. This Website will be maintained and updated by the authors, also providing errata and corrigenda of the code and possible mistakes in the book.

The authors wish to thank John Kimmel of Springer-Verlag and the referees for their valuable comments and publishing suggestions. In addition, they would like to acknowledge the University of Padova and the Italian Ministry for University and Scientific and Technological Research (MIUR - PRIN 2006) for providing the financial support for the necessary research and developing part of the R codes.

Dario Basso, Fortunato Pesarin, Luigi Salmaso, and Aldo Solari
Padova, December 2008

corresponding author: Luigi Salmaso, email: salmaso@gest.unipd.it
book Website: www.gest.unipd.it/~salmaso/web/springerbook.htm

This work has been supported by the Italian Ministry of University and Research (PRIN 2006133284_003) and by the University of Padova (CPDA088513). Research projects' coordinator: Prof. Luigi Salmaso.

Contents

Preface v

1 Permutation Tests 1
  1.1 Introduction 1
  1.2 Basic Construction 4
  1.3 Properties 7
  1.4 Multivariate Permutation Tests 10
      1.4.1 Properties of the Nonparametric Combination Tests ... 17
  1.5 Examples.. 18
      1.5.1 Univariate Permutation Tests 18
      1.5.2 The Nonparametric Combination Methodology 22
  1.6 Multiple Testing... 25
  1.7 Multiple Comparisons..................................... 32

Part I Stochastic Ordering

2 Ordinal Data 39
  2.1 Introduction 39
  2.2 Testing Whether Treatment is "Better" than Control 42
      2.2.1 Conditional Distribution 43
      2.2.2 Linear Test Statistics: Choice of Scores 44
      2.2.3 Applications with R functions...................... 51
      2.2.4 Concordance Monotonicity 53
      2.2.5 Applications with R functions...................... 55
      2.2.6 Multiple Testing 55
  2.3 Independent Binomial Samples 57
      2.3.1 Applications with R functions 60
  2.4 Comparison of Several Treatments when the Response is
      Ordinal 62

3   Multivariate Ordinal Data 65
  3.1   Introduction 65
  3.2   Standardized Test Statistics 72
  3.3   Multiple Testing on Endpoints and Domains 74
  3.4   Analysis of the FOB Data 76
  3.5   Violations of Stochastic Order 78

4   Multivariate Continuous Data 85
  4.1   Introduction 85
  4.2   Testing Superiority 86
  4.3   Testing Superiority and Noninferiority 93
    4.3.1   Applications with R functions 96
  4.4   Several Samples 97
    4.4.1   Applications with R functions 101

Part II Nonparametric ANOVA

5   Nonparametric One-Way ANOVA 105
  5.1   Overview of Nonparametric One-Way ANOVA 106
  5.2   Permutation Solution 107
    5.2.1   Synchronizing Permutations 110
    5.2.2   A Comparative Simulation Study for One-Way ANOVA 113
  5.3   Testing for Umbrella Alternatives 114
  5.4   Simple Stochastic Ordering Alternatives 116
  5.5   Permutation Test for Umbrella Alternatives 119
    5.5.1   The Mack and Wolfe Test 120
  5.6   A Comparative Simulation Study 122
  5.7   Applications with R 126
    5.7.1   One-Way ANOVA with R............................127
    5.7.2   Umbrella Alternatives with R 129

6   Synchronized Permutation Tests in Two-way ANOVA 133
  6.1   Introduction 133
  6.2   The Test Statistics 135
  6.3   Constrained and Unconstrained Synchronized Permutations...136
  6.4   Properties of the Synchronized Permutation Test Statistics ...140
    6.4.1   Uncorrelatedness Among Synchronized Permutation
        Tests 140
    6.4.2   Unbiasedness and Consistency of Synchronized
        Permutation Tests 143
  6.5   Power Simulation Study 146
  6.6   Multiple Comparisons 149
  6.7   Examples and Use of R Functions 154
    6.7.1   Applications with R Functions......................156

6.7.2 Examples 166
6.8 Further Developments.....................................168
6.8.1 Unbalanced Two-Way ANOVA Designs 168
6.8.2 Two-Way MANOVA 170

7 **Permutation Tests for Unreplicated Factorial Designs**......173
7.1 Brief Introduction to Unreplicated 2^K Full Factorial Designs ..174
7.2 Loughin and Noble's Test................................176
7.3 The T_F Test 180
7.4 The (Basso and Salmaso) T_P Test 184
7.5 The (Basso and Salmaso) Step-up T_P 186
7.5.1 Calibrating the Step-up T_p 192
7.6 A Comparative Simulation Study 195
7.7 Examples with R 198
7.7.1 Calibrating the Step-up T_P with R..................204

References..207

Index 215

1

Permutation Tests

1.1 Introduction

This book deals with the permutation approach to a variety of univariate and multivariate problems of hypothesis testing in a nonparametric framework. The great majority of univariate problems may be usefully and effectively solved within standard parametric or nonparametric methods as well, although in relatively mild conditions their permutation counterparts are generally asymptotically as good as the best parametric ones. Moreover, it should be noted that permutation methods are essentially of a nonparametrically exact nature in a conditional context. In addition, there are a number of parametric tests the distributional behavior of which is only known asymptotically. Thus, for most sample sizes of practical interest, the relative lack of efficiency of permutation solutions may sometimes be compensated by the lack of approximation of parametric asymptotic counterparts. Moreover, when responses are normally distributed and there are too many nuisance parameters to estimate and remove, due to the fact that each estimate implies a reduction of the degrees of freedom in the overall analysis, it is possible for the permutation solution to become better than its parametric counterpart (see, for example, Chapter 6). In addition, assumptions regarding the validity of parametric methods (such as normality and random sampling) are rarely satisfied in practice, so that consequent inferences, when not improper, are necessarily approximated, and their approximations are often difficult to assess.

For most problems of hypothesis testing, the observed data set $\mathbf{y} = \{y_1, \ldots, y_n\}$ is usually obtained by a symbolic experiment performed n times on a population variable Y, which takes values in the sample space $\mathcal{Y}$. We often add the adjective *symbolic* to names such as experiments, treatments, treatment effects, etc., in order to refer to experimental, pseudo-experimental, and observational contexts. For the purposes of analysis, the data set $\mathbf{y}$ is generally partitioned into *groups* or *samples*, according to the so-called *treatment levels* of the experiment. In the context of this chapter, we use capital letters

D. Basso et al., *Permutation Tests for Stochastic Ordering and ANOVA*, Lecture
Notes in Statistics, 194, DOI 10.1007/978-0-387-85956-9_1,
© Springer Science+Business Media, LLC 2009

for random variables and lower case letters for the observed data set. In some sections, we shall dispense with this distinction because the context is always sufficiently clear. Of course, when a data set is observed at its $\mathbf{y}$ value, it is presumed that a sampling experiment on a given underlying population has already been performed, so that the resulting sampling distribution is related to that of the parent population, which is usually denoted by P.

For any general testing problem, in the null hypothesis (H_0), which usually assumes that data come from only one (with respect to groups) unknown population distribution P, the whole set of observed data $\mathbf{y}$ is considered to be a random sample, taking values on sample space $\mathcal{Y}^n$, where $\mathbf{y}$ is one observation of the n-dimensional sampling variable $\mathbf{Y}^{(n)}$ and where this random sample does not necessarily have independent and identically distributed (i.i.d.) components. We note that the observed data set $\mathbf{y}$ is always a set of sufficient statistics in H_0 for any underlying distribution. In order to see this in a simple way, let us assume that H_0 is true and all members of a nonparametric family $\mathcal{P}$ of nondegenerate and distinct distributions are dominated by one *dominating* measure ξ; moreover, let us denote by f_P the density of P with respect to ξ, by $f_P^{(n)}(\mathbf{y})$ the density of the sampling variable $\mathbf{Y}^{(n)}$, and by $\mathbf{y}$ the data set. As the identity $f_P^{(n)}(\mathbf{y}) = f_P^{(n)}(\mathbf{y}) \cdot 1$ is true for all $\mathbf{y} \in \mathcal{Y}^n$, except for points such that $f_P^{(n)}(\mathbf{y}) = 0$, due to the well-known factorization theorem, any data set $\mathbf{y}$ is therefore a sufficient set of statistics for whatever $P \in \mathcal{P}$.

Note that a family of distributions $\mathcal{P}$ is said to behave nonparametrically when we are not able to find a parameter θ, belonging to a known finite-dimensional parameter space Θ, such that there is a one-to-one relationship between Θ and $\mathcal{P}$ in the sense that each member of $\mathcal{P}$ cannot be identified by only one member of Θ and vice versa.

By the *sufficiency*, *likelihood*, and *conditionality principles of inference* for a review, see Cox and Hinkley, 1974, Chapter 2), given a sample point $\mathbf{y}$, if $\mathbf{y}^* \in \mathcal{Y}^n$ is such that the likelihood ratio $f_P^{(n)}(\mathbf{y})/f_P^{(n)}(\mathbf{y}^*) = \rho(\mathbf{y}, \mathbf{y}^*)$ is not dependent on f_P for whatever $P \in \mathcal{P}$, then $\mathbf{y}$ and $\mathbf{y}^*$ are said to *contain essentially the same amount of information with respect to P*, so that they are equivalent for inferential purposes. The set of points that are equivalent to $\mathbf{y}$, with respect to the information contained, is called *the coset of* $\mathbf{y}$ or *the orbit associated with* $\mathbf{y}$, and is denoted by $\mathcal{Y}^n_{/\mathbf{y}}$, so that $\mathcal{Y}^n_{/\mathbf{y}} = \{\mathbf{y}^* : \rho(\mathbf{y}, \mathbf{y}^*) \text{ is } f_P\text{-independent}\}$. It should be noted that, when data are obtained by random sampling with i.i.d. observations, so that $f_P^{(n)}(\mathbf{y}) = \prod_{1 \le i \le n} f_P(y_i)$, the orbit $\mathcal{Y}^n_{/\mathbf{y}}$ associated with $\mathbf{y}$ contains all permutations of $\mathbf{y}$ and, in this framework, the likelihood ratio satisfies the equation $\rho(\mathbf{y}, \mathbf{y}^*) = 1$. Also note that, as in Chapter 6, orbits of f_P-invariant points may be constructed without permuting the whole data set.

The same conclusion is obtained if $f_P^{(n)}(\mathbf{y})$ is assumed to be invariant with respect to permutations of the arguments of $\mathbf{y}$; i.e., the elements $(y_1, \ldots, y_n)$. This happens when the assumption of independence for observable data is replaced by that of *exchangeability*, $f_P^{(n)}(y_1, \ldots, y_n) = f_P^{(n)}(y_{u_1^*}, \ldots, y_{u_n^*})$, where $(u_1^*, \ldots, u_n^*)$ is any permutation of $(1, \ldots, n)$. Note that, in the context of permutation tests, this concept of exchangeability is often referred to as the *exchangeability of the observed data with respect to groups*. Orbits $\mathcal{Y}_{/\mathbf{y}}^n$ are also called *permutation sample spaces*. It is important to note that orbits $\mathcal{Y}_{/\mathbf{y}}^n$ associated with data sets $\mathbf{y} \in \mathcal{Y}^n$ always contain a finite number of points, as n is finite.

Roughly speaking, permutation tests are conditional statistical procedures, where conditioning is with respect to the *orbit $\mathcal{Y}_{/\mathbf{y}}^n$ associated with the observed data set* $\mathbf{y}$. We will sometimes use use the notation $\Pr\{\cdot | \mathbf{y}\}$ instead of $\Pr\{\cdot | \mathcal{Y}_{/\mathbf{y}}\}$ to denote the conditioning with respect to the orbit associated with data set $\mathbf{y}$ even though the two notations are not necessarily equivalent. Thus, $\mathcal{Y}_{/\mathbf{y}}^n$ plays the role of *reference set for the conditional inference* (see Lehmann and Romano, 2005). In this way, in the null hypothesis and assuming exchangeability, the conditional probability distribution of a generic point $\mathbf{y}' \in \mathcal{Y}_{/\mathbf{y}}^n$, for any underlying population distribution $P \in \mathcal{P}$, is

$$\Pr\{\mathbf{y}^* = \mathbf{y}' | \mathcal{Y}_{/\mathbf{y}}^n\} = \frac{\sum_{\mathbf{y}^* = \mathbf{y}'} f_P^{(n)}(\mathbf{y}^*) \cdot d\xi^n}{\sum_{\mathbf{y}^* \in \mathcal{Y}_{/\mathbf{y}}^n} f_P^{(n)}(\mathbf{y}^*) \cdot d\xi^n} = \frac{\#[\mathbf{y}^* = \mathbf{y}', \ \mathbf{y}^* \in \mathcal{Y}_{/\mathbf{y}}^n]}{\#[\mathbf{y}^* \in \mathcal{Y}_{/\mathbf{y}}^n]},$$

which is P-independent. Of course, if there is only one point in $\mathcal{Y}_{/\mathbf{y}}^n$ whose coordinates coincide with those of $\mathbf{y}'$, (i.e., if there are no ties in the data set), and if permutations correspond to permutations of the arguments, then this conditional probability becomes $1/n!$. Thus, $\Pr\{\mathbf{y}^* = \mathbf{y}' | \mathcal{Y}_{/\mathbf{y}}^n\}$ is uniform on $\mathcal{Y}_{/\mathbf{y}}^n$ for all $P \in \mathcal{P}$.

These statements allow permutation inferences to be invariant with respect to P in H_0. Some authors, emphasizing this invariance property of permutation distribution in H_0, prefer to give them the name of *invariant tests*. However, due to this invariance property, permutation tests are distribution-free and nonparametric.

As a consequence, in the alternative hypothesis H_1, conditional probability shows quite different behavior and in particular may depend on P. To achieve this in a simple way, let us consider, for instance, a two-sample problem where $f_{P_1}^{(n_1)}$ and $f_{P_2}^{(n_2)}$ are the densities, relative to the same dominating measure ξ, of two sampling distributions related to two populations, P_1 and P_2, that are assumed to differ at least in a set of positive probability. Suppose also that $\mathbf{y}_1$ and $\mathbf{y}_2$ are the two separate and independent data sets with sample sizes n_1 and n_2, respectively. Therefore, as the likelihood associated with the pooled data set is $f_P^{(n)}(\mathbf{y}) = f_{P_1}^{(n_1)}(\mathbf{y}_1) \cdot f_{P_2}^{(n_2)}(\mathbf{y}_2)$, from the sufficiency principle it follows that the data set partitioned into two groups, $(\mathbf{y}_1; \mathbf{y}_2)$, is now the set of sufficient statistics. Indeed, by joint invariance of the likelihood ratio with respect to both f_{P_1} and f_{P_2}, the coset of $\mathbf{y}$ is $(\mathcal{Y}_{/\mathbf{y}_1}^{n_1}, \mathcal{Y}_{/\mathbf{y}_2}^{n_2})$, where $\mathcal{Y}_{/\mathbf{y}_1}^{n_1}$ and $\mathcal{Y}_{/\mathbf{y}_2}^{n_2}$ are partial orbits associated with $\mathbf{y}_1$ and $\mathbf{y}_2$, respectively. This implies that, conditionally, no datum from $\mathbf{y}_1$ can be exchanged with any other from $\mathbf{y}_2$ because in H_1 permutations are permitted only within groups, separately.

Consequently, when we are able to find statistics that are sensitive to the diversity of two distributions, we may have a procedure for constructing permutation tests. Of course, when constructing permutation tests, one should also take into consideration the physical meaning of treatment effects, so that the resulting inferential conclusions have clear interpretations.

Although the concept of conditioning for permutation tests is properly related to the formal conditioning with respect to orbit $\mathcal{Y}_{/\mathbf{y}}^n$, henceforth we shall generally adopt a simplified expression for this concept by stating that *permutation tests are inferential procedures that are conditional with respect to the observed data set* $\mathbf{y}$. Indeed, once $\mathbf{y}$ is known and the exchangeability condition is assumed in H_0, $\mathcal{Y}_{/\mathbf{y}}^n$ remains completely determined by $\mathbf{y}$.

1.2 Basic Construction

In this section, we provide examples on the construction of a permutation test. We will do this by considering a two-sample design. Let $\mathbf{y}_1$ and $\mathbf{y}_2$ be two independent samples of size n_1 and n_2 from two population distributions P_1 and P_2, respectively. In addition, let $P_1(y) = P_2(y - \delta)$. That is, the population distributions differ only in location. A common testing problem is to assess whether $P_1 \stackrel{d}{=} P_2$ or not, where the symbol $\stackrel{d}{=}$ means equality in distribution. In a location problem, there are several ways to specify the underlying model generating the observed data; for instance, let

$$
\begin{aligned}
Y_{i1} &= \mu_1 + \varepsilon_{i1}, & i &= 1, \ldots, n_1, \\
Y_{j2} &= \mu_1 + \delta + \varepsilon_{j2}, & j &= 1, \ldots, n_2,
\end{aligned}
\tag{1.1}
$$

be the models describing a generic observation from the first and second samples, respectively. Here $\delta = \mu_2 - \mu_1$, μ_1 and μ_2 are population constants, and ε_{i1} and ε_{j2} are identically distributed random variables with zero mean and variance $\sigma^2 < +\infty$ (the so-called experimental errors), not necessarily independent within the observations. The null hypothesis $P_1 \stackrel{d}{=} P_2$ can be written in terms of $\delta = 0$ against the alternative hypothesis $\delta \neq 0$. If H_0 is true, then Y_{i1} and Y_{j2} are identically distributued random variables. In addition, if ε_{i1} and ε_{j2} are exchangeable random variables, in the sense that $\Pr(\varepsilon) = \Pr(\varepsilon^*)$, where $\varepsilon = [\varepsilon_{11}, \varepsilon_{21}, \ldots, \varepsilon_{n_1 1}, \varepsilon_{12}, \varepsilon_{22}, \ldots, \varepsilon_{n_2 2}]'$ and ε^* is a permutation of ε, then also Y_{i1} and Y_{j2} are exchangeable in the sense that $\Pr(\mathbf{Y}) = \Pr(\mathbf{Y}^*)$, where $\mathbf{Y} = [\mathbf{Y_1}, \mathbf{Y_2}]'$ and $\mathbf{Y}^*$ is the corresponding permutation of $\mathbf{Y}$. As a simple example, consider the common case where the observations are independent. The likelihood can be written as

$$
\begin{aligned}
L(\delta; \mathbf{y}) &= f_{P_1}^{(n_1)}(y_{11}, y_{21}, \ldots, y_{n_1 1}) f_{P_2}^{(n_2)}(y_{12}, y_{22}, \ldots, y_{n_2 2}) \\
&= \prod_{i=1}^{n_1} f_{P_1}(y_{i1}) \prod_{j=1}^{n_2} f_{P_2}(y_{i2}),
\end{aligned}
$$

where $\mathbf{y} = [\mathbf{y}_1, \mathbf{y}_2]'$ and $f_{P_j}(y)$ is the density of Y_j, $j = 1, 2$. If H_0 is true, $f_{P_1}(y) = f_{P_2}(y)$, so $L_{H_0}(\delta; \mathbf{y}) = L_{H_0}(\delta; \mathbf{y}^*)$. Roughly speaking, this means that (conditionally), under H_0, $\mathbf{y}_1$ and $\mathbf{y}_2$ are two independent samples from the same population distribution P, or equivalently that $\mathbf{y}$ is a random sample of size $n = n_1 + n_2$ from P.

In order to obtain a statistical test, we need to define a proper test statistic and obtain its null distribution. How do we find the "best" test statistic for a given inferential problem? There is no specific answer to this question when the population distributions are unknown. One reasonable criterion is, for instance, to let the unconditional expectation of a chosen test statistic depend only on the parameter of interest. For instance, since unconditionally $E[\bar{y}_1] = \mu_1$ and $E[\bar{y}_2] = \mu_2$, a suitable test statistic could be defined as $T(\mathbf{y}) = \bar{y}_1 - \bar{y}_2$. Another reasonable choice is to look at the parametric counter-part: In a two-sample location problem, the well-known t statistic

$$t = \frac{\bar{y}_1 - \bar{y}_2}{\left[\left(\frac{1}{n_1} + \frac{1}{n_2}\right) s^2\right]^{\frac{1}{2}}}$$

can also be considered. We will see that the t statistic and $T(\mathbf{y}) = \bar{y}_1 - \bar{y}_2$ are equivalent within a permutation framework. By equivalent test statistics we mean test statistics that lead to the same rejection region in the permutation sample space $\mathcal{Y}^n/_{\mathbf{y}}$, so they also lead to the same inference for any given set $\mathbf{y} \in \mathcal{Y}$.

Within a permutation framework, a test statistic $T : \mathcal{Y}/_{\mathbf{y}} \to \mathcal{T}$ is a real function of all the observed data that takes values on the support $\mathcal{T} = T(\mathcal{Y}/_{\mathbf{y}}) \subseteq \mathbb{R}^1$. It is worth noting that the support $\mathcal{T}$ depends on $\mathbf{y}$ in the sense that whenever $\mathbf{y} \neq \mathbf{y}'$ we may have $\mathcal{T}_{\mathbf{y}} \neq \mathcal{T}_{\mathbf{y}'}$. Moreover, if T is such that $T(\mathbf{y}^{*\prime}) \neq T(\mathbf{y}^{*\prime\prime})$ for any two distinct points of $\mathcal{Y}/_{\mathbf{y}}$, in the null hypothesis the distribution of T over $\mathcal{T}$ is uniform; that, is all points are equally likely.

The null distribution of $T(\mathbf{y})$ is given by the elements of the space $\mathcal{T}$. We will use the notation $T(\mathbf{y})$, T°, or simply T to emphasize the observed value of the test statistic (the one obtained from the observed data), whereas T^* indicates a value of the permutation distribution of the test statistic. Note that $T^* = T^\circ$ if the identity permutation is applied to $\mathbf{y}$.

To perform a statistical test, we only need to define a distance function on $\mathcal{T}$ in order to specify which elements of $\mathcal{Y}/_{\mathbf{y}}$ are "far" from H_0. That is, we need a rule to determine the critical region of the test. To this end, let us explore the space $\mathcal{T}$ through the two-sample location problem example. Let $T(\mathbf{y}) = \bar{y}_1 - \bar{y}_2$ be the test statistic and $T^* = \bar{y}_1^* - \bar{y}_2^*$ be the generic element of $\mathcal{T}$. Conditionally, the expectation and variance of observations in $\mathbf{y}_1$ and $\mathbf{y}_2$ are, respectively

$$E(y_{i1}|\mathbf{y}) = \bar{y}_1, \qquad E(y_{i1}^2|\mathbf{y}) = \frac{1}{n_1}\sum_i y_{i1}^2,$$

$$E(y_{j2}|\mathbf{y}) = \bar{y}_2, \qquad E(y_{j2}^2|\mathbf{y}) = \frac{1}{n_2}\sum_j y_{y2}^2.$$

Now let y_{i1}^* be a generic observation in $\mathbf{y}_1^*$. Conditionally, $\Pr[y_{i1}^* \in \mathbf{y}_1|\mathbf{y}] = n_1/n$ and $\Pr[y_{i1}^* \in \mathbf{y}_2|\mathbf{y}] = n_2/n$. The conditional expected value of y_{i1}^* is therefore

$$E[y_{i1}^*|\mathbf{y}] = \frac{n_1}{n}\bar{y}_1 + \frac{n_2}{n}\bar{y}_2 = \bar{y}.$$

Similarly, $E[y_{j2}^*|\mathbf{y}] = \bar{y}$. Consequently, $E[T^*|\mathbf{y}] = 0$, and therefore the null distribution of T^* is centered, although it is not necessarily symmetric, in the sense that $F_{T^*}(t^*) = 1 - F_{T^*}(-t^*)$, $t^* \in \mathcal{T}$. It is symmetric, for instance, in the balanced case where $n_1 = n_2$. As regards the variance

$$\mathrm{Var}(y_{i1}^*|\mathbf{y}) = E[y_{i1}^{*2}|\mathbf{y}] - E[y_{i1}^*|\mathbf{y}]^2 = \frac{1}{n}\sum_{l=1}^{2}\sum_{k=1}^{n_l} y_{kl}^2 - \bar{y}^2 = \hat{\sigma}_0^2,$$

where $\hat{\sigma}_0^2$ is the maximum likelihood estimate of the variance under H_0 when data are normally distributed. Note that $\hat{\sigma}_0^2$ is constant, in a conditional framework. Note also that the y_{i1}^*'s are not independent. By the finite population theory,

$$\mathrm{Var}(\bar{y}_1^*|\mathbf{y}) = \frac{\hat{\sigma}_0^2}{n_1}\left(\frac{n-n_1}{n-1}\right) = \frac{\hat{\sigma}_0^2}{n-1}\frac{n_2}{n_1}.$$

Now consider the relationship $n_1\bar{y}_1^* + n_2\bar{y}_2^* = Y$, where Y is the total of observations, which is permutationally invariant. Then

$$\mathrm{Var}(T^*|\mathbf{y}) = \mathrm{Var}\left(\bar{y}_1^* - \frac{Y}{n_2} + \frac{n_1\bar{y}_1^*}{n_2}\bigg|\mathbf{y}\right) = \mathrm{Var}\left(\frac{n}{n_2}\bar{y}_1^*\bigg|\mathbf{y}\right) = \frac{n^2}{n_2^2}\mathrm{Var}(\bar{y}_1^*|\mathbf{y})$$

$$= \frac{n\hat{\sigma}_0^2}{n-1}\frac{n}{n_1 n_2} = \frac{n\hat{\sigma}_0^2}{n-1}\left(\frac{n_1+n_2}{n_1 n_2}\right) = s_0^2\left(\frac{1}{n_1}+\frac{1}{n_2}\right),$$

which is like the denominator of the t test, despite the estimate of the population variance. Note that s_0^2 is the unbiased estimate of $\mathrm{Var}(Y)$ when H_0 is true. Therefore, we can define a test statistic as

$$T^* = \frac{n_1 n_2}{n}\frac{(\bar{y}_1^* - \bar{y}_2^*)^2}{s_0^2}, \tag{1.2}$$

where the emphasis is on the fact that T^* is a random variable defined on $\mathcal{Y}/_\mathbf{y}$. Large values of (1.2) are significant against the null hypothesis. Since n_1, n_2 and s_0^2 are constant, (1.2) is permutationally equivalent to $T^{*\prime} = (\bar{y}_1^* - \bar{y}_2^*)^2$ and to $T^{*\prime\prime} = |\bar{y}_1^* - \bar{y}_2^*|$.

A similar proof applies to the classic t statistic: Let t^{*2} be the (squared) value of the t statistic obtained from a random permutation of $\mathbf{y}^*$,

$$t^{*2} = \frac{n_1 n_2}{n} \frac{(\bar{y}_1^* - \bar{y}_2^*)^2}{s^{*2}}.$$

It can be easily proved (see Section 5.2) that this is a special case of one-way ANOVA framework (when $C = 2$). Therefore, t^{*2} is a monotone nondecreasing function of $T^{*\prime}$, and since permutation tests are based on the ordered values of $\mathcal{T}$ (see Section 1.3), t^{*2} is permutationally equivalent to T^* as well.

The exact p-value of the test is

$$p = \frac{1}{C} \sum_{T^* \in \mathcal{T}} I(T^* \geq T^o) = \frac{\#[T^* \geq T^o]}{C},$$

where $T^o = T(\mathbf{y})$, $I(\cdot)$ is the indicator function, and C is the cardinality of $\mathcal{T}$. If Y is a continuous random variable (i.e., the probability of having ties is zero), then

$$C = \binom{n}{n_1}.$$

Clearly C increases very rapidly with n, so in practice the c.d.f. of T^* is approximated by a Monte Carlo sampling from $\mathcal{T}$. Let B be the number of Monte Carlo permutations. Then the c.d.f. of T^* is estimated by

$$\hat{F}_{T^*}(t) = \frac{\#[T^* \leq t]}{B} \qquad t \in \mathbb{R}.$$

1.3 Properties

In this section, we investigate some properties of the permutation tests, such as exactness and unbiasedness; for consistency we refer to Hoeffding (1952). Let Y be a random variable such that $E[Y] = \mu$ and $Var[Y]$ exists. Let $H_0 : \mu \leq \mu_0$ be the null hypothesis to be assessed and $T = T(\mathbf{Y})$ a suitable test statistic for H_0 (in the sense that large values of T are significant against H_0). Then, a (nonrandomized) test ϕ of size α is a function of the test statistic $T = T(Y)$ such as

$$\phi(T) = \begin{cases} 1 & \text{if} \quad T \geq T^{1-\alpha} \\ 0 & \text{if} \quad T < T^{1-\alpha}, \end{cases}$$

where $T^{1-\alpha}$ is the $1 - \alpha$ quantile of the null distribution of T, i.e. $\Pr[T \geq T^{1-\alpha}|\mathcal{Y}_{/\mathbf{y}}] = \alpha$. The α-values that satisfy $\Pr[T \geq T^{1-\alpha}|\mathcal{Y}_{/\mathbf{y}}] = \alpha$ are called attainable α-values. The set of attainable α-values is a proper subset of $(0, 1]$. Thus, if $H_0 : \mu = \mu_0$, permutation tests are exact for all attainable α-values, whereas if $H_0 : \mu \leq \mu_0$, they are conservative.

If the distribution of T is symmetric, one can define a test for two-sided alternatives by replacing T with $|T|$ in the definition of ϕ, or T^α with $T^{1-\alpha}$ if the alternative hypothesis is $H_1 : \mu < \mu_0$. Clearly, the expected value of ϕ is

$$E[\phi] = 1 \cdot \Pr[T \geq T^{\alpha}] + 0 \cdot \Pr[T < T^{\alpha}] = \alpha.$$

That is why ϕ is usually called a test of size α.

Permutation tests are conditional procedures; therefore the definitions of the usual properties of *exactness* and *unbiasedness*, and *consistency* require an ad hoc notation: From now on, we denote by $\mathbf{y}(\delta)$ the set of data when the alternative hypothesis is true and by $\mathbf{y}(\mathbf{0})$ the set of data when the null hypothesis is true.

The test ϕ of size α is said to be exact if $\forall\, 0 < \alpha < 1$:

$$\Pr[\phi = 1|\mathbf{y}(\mathbf{0})] = \alpha.$$

The test ϕ is said to be unbiased if

$$\Pr[\phi = 1|\mathbf{y}(\mathbf{0})] \leq \alpha \leq \Pr[\phi = 1|\mathbf{y}(\delta)].$$

The test ϕ is said to be consistent if

$$\lim_{n \to +\infty} \Pr[\phi = 1|\mathbf{y}(\delta)] = 1.$$

To prove the properties of permutation tests, we will still refer to a univariate two-sample problem. In the previous section, we have given an informal definition of a permutation test.

Formally, let $\mathcal{Y}^n/\mathbf{y}$ be the orbit associated with the observed vector of data $\mathbf{y}$. The points of $\mathcal{Y}^n/\mathbf{y}$ can also be defined as $\mathbf{y}^* : \mathbf{y}^* = \pi\mathbf{y}$ where π is a random permutation of indexes $1, 2, \ldots, n$. Define a suitable test statistic T on $\mathcal{Y}^n/\mathbf{y}$ for which large values are significant for a right-handed one-sided alternative: The image of $\mathcal{Y}^n/\mathbf{y}$ through T is the set $\mathcal{T}$ that consists of C elements (if there are no ties in the given data). Let

$$T^*_{(1)} \leq T^*_{(2)} \leq \cdots \leq T^*_{(C)}$$

be the ordered values of $\mathcal{T}$. Let T^o be the observed value of the test statistic, $T^o = T(\mathbf{y})$. For a chosen attainable significance level $\alpha \in \{1/C, 2/C, \ldots, (C-1)/C\}$, let $k = C(1-\alpha)$. Define a permutation test for a one-sided alternative the function $\phi^* = \phi(T^*)$

$$\phi^*(T) = \begin{cases} 1 & \text{if} & T^o \geq T^*_{(k)} \\ 0 & \text{if} & T^o < T^*_{(k)} \end{cases}.$$

Since the critical values of the distribution of T^* depend on the observed data, one can provide a more general definition of a permutation test based on the p-values, whose distribution depends on sample size n:

$$\phi^*(T) = \begin{cases} 1 & \text{if} & \Pr[T^* \geq T^o|\mathbf{y}] \leq \alpha \\ 0 & \text{if} & \Pr[T^* \geq T^o|\mathbf{y}] > \alpha \end{cases}.$$

The equivalence of the two definitions is ensured by the relationship

$$\Pr\{\Pr[T^* \geq T^o|\mathbf{y}] \leq \alpha\} = \Pr\{\Pr[T^* \leq T^o|\mathbf{y}] \geq 1 - \alpha\}$$
$$= \Pr\{F_{T^*}(T^o) \geq 1 - \alpha\}$$
$$= \Pr\{F_{T^*}^{-1}(F_{T^*}(T^o)) \geq F_{T^*}^{-1}(1 - \alpha)\}$$
$$= \Pr\{T^o \geq T_{(k)}^*\}.$$

To prove exactness, suppose H_0 is true. Then the elements of $\mathcal{T}$ are equally likely under the null hypothesis. This means that

$$\Pr\{T^* = T^o|\mathbf{y}(0)\} = \frac{1}{C} \quad \Rightarrow \quad \Pr\{T^o \in \mathcal{A}|\mathbf{y}(0)\} = \frac{\#[T^* \in \mathcal{A}]}{C},$$

where $\mathcal{A}$ is one element of the Borel set defined on $\mathcal{T}$. Hence, for any attainable significance level α

$$\Pr\{\phi^*(T) = 1|\mathbf{y}(0)\} = \Pr\{T^o \geq T_{(k)}^*|\mathbf{y}(0)\}$$
$$= \frac{\#[T^* \geq T_{(k)}^*]}{C} = \frac{C\alpha}{C} = \alpha.$$

Note that, since permutation tests are conditionally exact, they are unconditionally exact as well.

As regards unbiasedness, we will refer to the two-sample problem of the previous section. Let's suppose that data of the two samples are generated under the model (1.1), and let $H_0 : \mu_2 - \mu_1 \leq 0$ be the null hypothesis to assess. Define the test statistic as $T = \bar{y}_2 - \bar{y}_1$. Let $T^o(0)$ and $T^o(\delta)$ be respectively the observed value of T when data are $\mathbf{y}(0)$ and $\mathbf{y}(\delta)$, respectively,

$$T^o(0) = T^*(\mathbf{y}(0)) : \bar{y}_2 - \bar{y}_1 = \bar{\varepsilon}_2 - \bar{\varepsilon}_1,$$
$$T^o(\delta) = T^*(\mathbf{y}(\delta)) : \bar{y}_2 - \bar{y}_1 = \delta + \bar{\varepsilon}_2 - \bar{\varepsilon}_1,$$

where $\bar{\varepsilon}_2$ and $\bar{\varepsilon}_1$ are sampling averages of n_2 and n_1 exchangeable errors, respectively. Since the event $\Pr[T^* \geq T^o|\mathcal{Y}/\mathbf{y}] \leq \alpha$ implies the event $\{T^o \geq T_{(k)}^*|\mathcal{Y}/\mathbf{y}\}$, we may write

$$\Pr[T^o(0) \geq T_{(k)}^*|\mathbf{y}(0)] = \Pr[\bar{\varepsilon}_2 - \bar{\varepsilon}_1 \geq T_{(k)}^*]$$

and

$$\Pr[T^o(\delta) \geq T_{(k)}^*|\mathbf{y}(\delta)] = \Pr[\bar{\varepsilon}_2 - \bar{\varepsilon}_1 \geq T_{(k)}^* - \delta].$$

Now, without loss of generality, let $\delta \geq 0$ and $T_{(k)}^* \geq 0$. Then, from the exactness of ϕ^*, we have:

$$\Pr[\phi^* = 1|\mathbf{y}(\delta)] = \Pr[T^o \geq T_{(k)}^*|\mathbf{y}(\delta)] \geq \Pr[T^o \geq T_{(k)}^*|\mathbf{y}(0)] = \Pr[\phi^* = 1|\mathbf{y}(0)],$$

which proves unbiasedness.

1.4 Multivariate Permutation Tests

There are some problems where the complexity requires a further approach. Consider, for instance, a multivariate problem where q (possibly dependent) variables are considered, or a multiaspect problem (such as the Beherens-Fisher problem), or a stratified analysis. The difficulties arise because of the underlying dependence structure among variables (or aspects), which is generally unknown. Moreover, a global answer involving several dependent variables (aspects) is often required, so the question is how to combine the information related to the q variables (aspects) into one global test.

Let us consider a one-sample multivariate problem with q dependent variables: Here the data set $\mathbf{Y}$ is an $n \times q$ matrix, where n is the sample size. What we are generally interested in is to test the null hypothsis $H_0 : \boldsymbol{\mu} = \boldsymbol{\mu_0}$ against the alternative hypothesis $H_1 : \boldsymbol{\mu} \neq \boldsymbol{\mu_0}$, where $\boldsymbol{\mu}$ is a $q \times 1$ vector of population means and $\boldsymbol{\mu_0} = [\mu_{01}, \mu_{02}, \ldots, \mu_{0q}]$ is a target vector. Assuming $\mathbf{Y}_i$ $i = 1, \ldots, n$ is a multivariate normal random variable, a parametric solution is Hotelling's T^2 test. In a bivariate problem, we may specify it as

$$
\begin{aligned}
T^2 &= n[\bar{\mathbf{y}} - \boldsymbol{\mu_0}]' \boldsymbol{\Sigma}^{-1} [\bar{\mathbf{y}} - \boldsymbol{\mu_0}] \\
&= \frac{n[\bar{y}_1 - \mu_1]^2}{s_1^2(1 - \hat{\rho}_{12}^2)} + \frac{n[\bar{y}_2 - \mu_2]^2}{s_2^2(1 - \hat{\rho}_{12}^2)} - 2\frac{n\hat{\rho}_{12}[\bar{y}_1 - \mu_1][\bar{y}_2 - \mu_2]}{s_1^2 s_2^2 (1 - \hat{\rho}_{12}^2)} \\
&= T(\mathbf{x_1}, \mu_1 | \hat{\rho}_{12}) + T(\mathbf{x_2}, \mu_2 | \hat{\rho}_{12}) - 2T'(\mathbf{x_1}, \mathbf{x_2}, \mu_1, \mu_2),
\end{aligned}
$$

where $T(\cdot)$ and $T'(\cdot)$ are test statistics, $\hat{\rho}_{12}$ is the estimate of the correlation between Y_1 and Y_2, and s_1^2 and s_2^2 are unbiased estimates of population variances. Note that Hotelling's T^2 is a combination of marginal tests on μ_1 and μ_2 accounting for the dependence between Y_1 and Y_2. Hotelling's T^2 depends on the estimated variance-covariance matrix $\boldsymbol{\Sigma}$, which has rank $n - q$, and it is appropriate only for two-sided alternatives. This means that either when $n \leq q$ or alternatives are one-sided, the Hotelling T^2 test cannot be applied. If Y_1 and Y_2 are independent, Hotelling's T^2 reduces to

$$
T^2 = \frac{n[\bar{y}_1 - \mu_1]^2}{s_1^2} + \frac{n[\bar{y}_2 - \mu_2]^2}{s_2^2} = T(\mathbf{y_1}, \mu_1 | \rho_{12} = 0) + T(\mathbf{y_2}, \mu_2 | \rho_{12} = 0).
$$

Within a conditional approach, there are no assumptions on the dependence structure among the q variables. Let us consider the matrix of observations partitioned into n q-dimensional arrays; that is,

$$
\mathbf{Y_{n \times q}} = \begin{bmatrix} y_{11} & y_{12} & \cdots & y_{1q} \\ y_{21} & y_{22} & \cdots & y_{2q} \\ \vdots & \vdots & \ddots & \vdots \\ y_{n1} & y_{n2} & \cdots & y_{nq} \end{bmatrix}.
$$

Each row of $\mathbf{Y}$ is a determination of the multivariate variable $[Y_1, Y_2, \ldots, Y_q]$, which has distribution P with unknown dependence structure. But, being determinations of the same random variable, the rows of $\mathbf{Y}$ (i.e., the data related to the statistical units) have an intrinsic dependence structure, which does not need to be modelled in order to do a permutation test if the permutations involve the rows of $\mathbf{Y}$. Note that this is true even if the vectors of the observations are repeated measures, or functions of the same data (e.g., the first r powers of a random variable Y).

A suitable nonparametric test to assess the hypothesis on marginal distributions $H_{0j} : \mu_j = \mu_{0j}, j = 1, \ldots, q$, is McNemar's test,

$$S_j = \sum_{i=1}^{n} I(y_{ij} - \mu_{0j} > 0),$$

where $I(\cdot)$ is the indicator function. If data in $\mathbf{Y_j}$ are symmetric and H_{0j} is true, then μ_{0j} represents the mean and the median of the distribution. Therefore, if μ_{0j} is true, S_j should be close to $n/2$. The null distribution of S_j is binomial with parameters n and $1/2$. Clearly, S_j is significant for small and large values, and the p-value of the test is obtained as

$$p_j = \Pr[X \leq (n - S_j)] + \Pr[X \geq S_j] \qquad \text{where} \qquad X \sim \text{Bi}(n, 1/2).$$

An equivalent version of McNemar's test is the test statistic

$$T^*(\mathbf{y_j}, \mu_{0j}) = \sum_{i=1}^{n} (y_{ij} - \mu_{0j})\text{sgn}^*(y_{ij} - \mu_{0j}), \qquad (1.3)$$

where

$$\Pr[\text{sgn}^*(y_i - \mu_0) = z] = \begin{cases} 1/2 & \text{if } z = +1 \\ 1/2 & \text{if } z = -1 \end{cases}.$$

Note that in one-sample location problems, the usual permutations do not apply since what is really informative here on the location parameter is the vector of observed signs $\mathbf{S_j} = [I(y_{1j} - \mu_{0j} > 0), I(y_{2j} - \mu_{0j} > 0), \ldots, I(y_{nj} - \mu_{0j} > 0)]$. According to McNemar's test, two points $\mathbf{y_j^*}$ and $\mathbf{y_j'}$ have the same likelihood if $\sum_{i=1}^{n} I(y_{ij}^* - \mu_{0j} > 0) = \sum_{i=1}^{n} I(y_i' - \mu_{0j} > 0)$. Here, the permutation sample space $\mathcal{Y}^{(n)}/_{\mathbf{y_j}}$ is given by

$$\mathcal{Y}^{(n)}/_{\mathbf{y_j}} = \{\mathbf{y_j^*} : \mathbf{y_j^*} = \pi^\pm(\mathbf{y_j} - \mu_{0j})\},$$

where $\pi^\pm$ is a combination of $n \pm$ signs, $\mu_{0j} = \mu_{0j}\mathbf{1_n}$ and $\mathbf{1_n}$ is an $n \times 1$ vector of 1's. The permutation sample space therefore has 2^n points. Note that in (1.3) we have

$$E[T^*(\mathbf{y_j}, \mu_{0j})|\mathbf{y_j}] = 0,$$

$$\text{Var}[T^*(\mathbf{y_j}, \mu_{0j})|\mathbf{y_j}] = \sum_{i=1}^{n} (y_{ij} - \mu_{0j})^2,$$

so the null distribution is always centered on μ_{0j}.

Since we have the relationship

$$H_0 : \boldsymbol{\mu} = \boldsymbol{\mu_0} \quad \Longrightarrow \quad \bigcap_{j=1}^{q} H_{0j},$$

the *global* null hypothesis H_0 can be viewed as an intersection of *partial* null hypotheses H_{0j}. Let λ_j, $j = 1, \ldots, q$, be a *partial* test statistic for the univariate hypothesis H_{0j}. By partial test we mean a test statistic to assess H_{0j} : $\mu_j = \mu_{0j}$ $j = 1, \ldots, q$. For instance, one may consider $\lambda_j = |T_j^*(\mathbf{y_j}, \mu_{0j})|$ or $\lambda_j = T_j^*(\mathbf{y_j}, \mu_{0j})^2$, which is significant for large values against $H_{0j} : \mu_j = \mu_{0j}$. The partial test statistics may also be significant for one-sided alternatives. For instance if $H_{1j} : \mu_j < \mu_{0j}$, then a test statistic is $\lambda_j = -T_j^*(\mathbf{y_j}, \mu_{0j})$. Now let

$$\psi^* = \psi(\mathbf{Y}^*, \boldsymbol{\mu_0}) = \sum_{j=1}^{q} \lambda_j \tag{1.4}$$

be a *global test statistic*. In order to account for the (possible) dependence among the q variables, the domain of ψ^* is

$$\mathcal{Y}^{(n)}/_{\mathbf{Y}^*} = \left\{ \mathbf{Y}^* : \mathbf{Y}^* = [\pi^{\pm}(\mathbf{y}_1 - \boldsymbol{\mu_{01}}), \pi^{\pm}(\mathbf{y}_2 - \boldsymbol{\mu_{02}}), \ldots, \pi^{\pm}(\mathbf{y}_q - \boldsymbol{\mu_{0q}})] \right\},$$

where $\pi^{\pm}$ is the same combination of $n \pm$ signs applied to all q vectors. If the q variables are independent, one may consider

$$\mathcal{Y}^{(n)}/_{\mathbf{Y}_{\perp}} = \left\{ \mathbf{Y}_{\perp}^* : \mathbf{Y}_{\perp}^* = [\pi_1^{\pm}(\mathbf{y}_1 - \boldsymbol{\mu_{01}}), \pi_2^{\pm}(\mathbf{y}_2 - \boldsymbol{\mu_{02}}), \ldots, \pi_q^{\pm}(\mathbf{y}_q - \boldsymbol{\mu_{0q}})] \right\}.$$

where the $\pi_i^{\pm}$'s are q independent combinations of $n \pm$ signs.

Note that $\mathcal{Y}^{(n)}/_{\mathbf{Y}_{\perp}}$ and $\mathcal{Y}^{(n)}/_{\mathbf{Y}^*}$ are different spaces. In particular, $\mathcal{Y}^{(n)}/_{\mathbf{Y}^*} \subseteq \mathcal{Y}^{(n)}/_{\mathbf{Y}_{\perp}}$, where $\mathcal{Y}^{(n)}/_{\mathbf{Y}_{\perp}}$ is the orbit associated to $\mathbf{Y}$ if the q variables are assumed to be independent, whereas in $\mathcal{Y}^{(n)}/_{\mathbf{Y}^*}$ the inner dependence among variables is maintained. The cardinality of $\mathcal{Y}^{(n)}/_{\mathbf{Y}_{\perp}}$ is 2^{nq}, whereas the cardinality of $\mathcal{Y}^{(n)}/_{\mathbf{Y}^*}$ is 2^n since the same combinations of signs apply to all q vectors.

If (1.5) is computed on $\mathcal{Y}^{(n)}/_{\mathbf{Y}^*}$, then $T^*(\mathbf{y_j}, \mu_{0j}) = T^*(\mathbf{y_j}, \mu_{0j}|\boldsymbol{\Sigma})$, where $\boldsymbol{\Sigma}$ is the matrix of (true) variances and covariances among q variables. That is, since the test statistic is defined on a permutation sample space accounting for dependence, the partial test statistic T_j^*'s also account for dependence. If $q = 2$, then let

$$H_0^G = \begin{cases} H_{01} : \mu_1 \le 0 \\ H_{02} : \mu_2 \le 0 \end{cases}$$

be the global null hypothesis, which is true if the partial null hypotheses H_{01} and H_{02} are jointly true and which should be rejected whenever one of the partial null hypotheses is rejected. Define a global test to assess H_0^G as

$$\psi^* = \psi(\mathbf{Y}^*, \mathbf{0}) = T_1^*(\mathbf{y_1}, 0) + T_2^*(\mathbf{y_2}, 0), \tag{1.5}$$

where $\mathbf{0} = [0,0]'$. The test statistic (1.5) is a direct combination of two partial tests for one-sided alternatives. Clearly, large values of ψ^* are significant against the global null hypothesis.

We may define

$$\mathcal{Y}^{(n)}/_{\mathbf{Y}*} = \{\mathbf{Y}^* : \mathbf{Y}^* = [\mathbf{y_1}, \mathbf{y_2}]^*\}$$

and

$$\mathcal{Y}^{(n)}/_{\mathbf{Y}_\perp} = \{\mathbf{Y}_\perp : \mathbf{Y}_\perp = [\mathbf{y_1}^*, \mathbf{y_2}^*]\},$$

where the emphasis is on the fact that in $\mathcal{Y}^{(n)}/_{\mathbf{Y}*}$ the permutations of signs involve the two columns of $\mathbf{Y}$ simultaneously, whereas in $\mathcal{Y}^{(n)}/_{\mathbf{Y}_\perp}$ the permutations of signs are done independently in each column of $\mathbf{Y}$.

Figure 1.1 represents the space $\mathcal{T} = \{[T_1^*, T_2^*] : T_j^* = T^*(\mathbf{y_j}^*, 0), j = 1, 2\}$ when $\mathbf{y_j^*} \in \mathcal{Y}^{(n)}/_{\mathbf{Y}_\perp}$ (black dots) and when $\mathbf{y_j^*} \in \mathcal{Y}^{(n)}/_{\mathbf{Y}*}$ (white dots) in a bivariate problem with $n = 4$ observations. Data have been generated from a bivariate normal distribution with

$$\mu = [0,1] \quad \text{and} \quad \Sigma = \begin{bmatrix} 1 & 0.5 \\ 0.5 & 1 \end{bmatrix}.$$

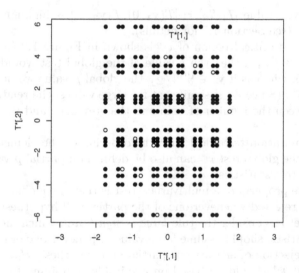

Fig. 1.1. Representation of $\mathcal{T}$ (black dots) and $\mathcal{T}_\perp$ (white dots) in a bivariate problem with $n = 4$.

It is evident how the dependence between Y_1 and Y_2 is reflected on the space $\mathcal{T}$ when $\mathbf{y_j^*} \in \mathcal{Y}^{(n)}/_{\mathbf{Y}*}$ (look at the white dots), which has 16 elements. On the other hand, if the domain of ψ^* is the space $\mathcal{Y}^{(n)}/_{\mathbf{Y}_\perp}$ (which consists of 256 points), then the random variable T_1^* is clearly independent of T_2^*.

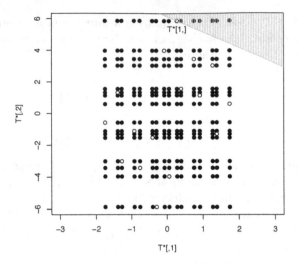

Fig. 1.2. Representation of $\mathcal{T}$ (black dots) and $\mathcal{T}_\perp$ (white dots) in a bivariate problem with $n = 4$. The critical region observed is given by the points in the grey zone.

The observed value $[T_1, T_2] = [T(\mathbf{y_1}, 0), T(\mathbf{y_2}, 0)]$ is highlighted with the symbol $T^*[1,]$ (see Section 1.5 for details).

The observed critical region of ψ^* is shown in Figure 1.2 (highlighted in grey). If $\mathbf{y}^* \in \mathcal{Y}^{(n)}/_{\mathbf{Y}^*}$, then the p-value of the global test would be equal to $1/16 = 0.0625$, whereas if $\mathbf{y}^* \in \mathcal{Y}^{(n)}/_{\mathbf{Y}_\perp}$ the global p-value would be equal to $7/256 = 0.0273$. (Since ψ^* is significant for large values, the related p-value is the ratio between the number of points in the grey zone and the total points considered).

Since permutation tests deal with the ordering of the elements of $\mathcal{T}$ (see Section 1.3), the global test ψ^* can also be defined on partial p-values instead of the partial test statistics.

A desirable property of a multivariate test is that the global null hypotheses should be rejected whenever one of the partial null hypothesis is rejected. To this end, let us consider the rule *large is significant*, which means that the global test statistic should assume large values whenever one of its arguments leads to the rejection of at least one partial null hypothesis H_{0j}. Accordingly, the global test ψ^* should be based on a suitable combining function ψ that satisfies the following requirements:

1. ψ must be continuous in all its q arguments.
2. ψ must be nondecreasing in its arguments. By this we mean that

$$\psi(\lambda_1, \ldots, \lambda_j, \ldots, \lambda_q) \geq \psi(\lambda_1, \ldots, \lambda'_j, \ldots, \lambda_q)$$

if λ_j is more significant against H_{0j} than λ'_j.

3. ψ must reach its supremum (possibly not finite) when one of its arguments tends to reject the related partial null hypothesis with probability one; that is,

$$\psi(\lambda_1, \ldots, \lambda_j, \ldots, \lambda_q) \to +\infty$$

if λ_j is "extremely" significant against H_{0j}. The meaning of the word "extremely" will be clearer in what follows.

The λ's in the definition of the combining function could be either test statistics or p-values. For instance, if the λ's are test statistics that are significant for large values (as in the bivariate example), some suitable combining functions are the following:

- the direct combining function: $\psi = \sum_{j=1}^{q} T_j$;
- the max T combining function: $\psi = \max_j T_j$.

Instead, if the combining function is based on the partial p-values (i.e., $\lambda_j = p_j = \Pr[T_j^* \geq T_j | \mathbf{Y}]$, which are significant against H_{0j} for small values), the following combining functions are of interest:

- Fisher's: $\psi = -2 \sum_{j=1}^{q} \log(p_j)$, $0 < \psi < +\infty$;
- Tippett's: $\psi = 1 - \min_j p_j$, $0 \leq \psi \leq 1$;
- Liptak's: $\psi = \sum_{j=1}^{q} \phi^{-1}(1 - p_j)$, where ϕ is the standard normal cumulative distribution function, $-\infty < \psi < +\infty$.

Property (3) of ψ means that if $\lambda_j = T_j$, then $\psi \to +\infty$ whenever $T_j \to +\infty$. On the other hand, if $\lambda_j = p_j$, then $\psi \to +\infty$ whenever $p_j \xrightarrow{p} 0$.

The combining functions above are particular cases of the nonparametric combination (NPC) of dependent tests introduced by Pesarin (2001). Now let's consider a global test when the global test statistic is defined on partial p-values; i.e., when the arguments of the combining function are $\lambda_j = \Pr[T_j^* \geq T_j | \mathbf{y_j}]$, $j = 1, 2$. Figure 1.3 represents the permutation space of the partial p-values that have been obtained from the spaces $\mathcal{T}$ (white dots) and $\mathcal{T}_\perp$ (black dots). The point labeled with the symbol $\mathbf{p[1,]}$ is the point whose coordinates are the p-values related to the observed statistics T_1 and T_2. In the bivariate example, $T_1 = 0.28907$ and $T_2 = 5.86235$. The related p-values are respectively $p_1 = \Pr[T_1^* \geq T_1 | \mathbf{y_1}] = 7/16$ and $p_2 = \Pr[T_2^* \geq T_2 | \mathbf{y_2}] = 1/16$. The remaining points of Figure 1.3 have coordinates equal to

$$[p_1^*, p_2^*] = [\Pr\{T_1^* \geq t_1^* | \mathbf{y_1}\}, \Pr\{T_2^* \geq t_2^* | \mathbf{y_2}\}] \quad [t_1^*, t_2^*] \in \mathcal{T}_\perp;$$

i.e., the generic point $[p_1^*, p_2^*]$ represents the p-values for each one-sided testing problem as if the observed values of the test statistics were the pair $[t_1^*, t_2^*]$ instead of $[T_1, T_2]$. If Fisher's combining function is applied to p-values, then $\psi^* = -2[\log(p_1^*) + \log(p_2^*)]$ gives the permutation null distribution of ψ. The global test statistic ψ is significant for large values, which are observed whenever one of its arguments tends to its infimum (here $\min(p_j) = 1/16$).

The critical region of the combining function ψ is indicated by the grey zone of Figure 1.3. The points that lay in this region are at least as significant

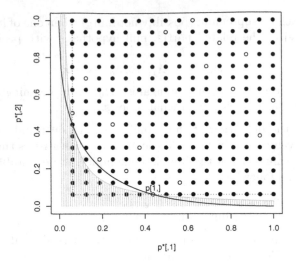

Fig. 1.3. Representation of $[\Pr(T_1^* \geq T_1^o), \Pr(T_2^* \geq T_2^o)]$ when $[T_1^*, T_2^*] \in \mathcal{T}$ (black dots) and when $[T_1^*, T_2^*] \in \mathcal{T}_\perp$ (white dots) in a bivariate problem with $n = 4$. The curves represent the border of the observed critical regions when different combining functions are applied.

as $[T_1, T_2]$ against the global null hypothesis $H_0^G = \cap_{j=1}^2 H_{0j}$ (where $H_{0j} : \mu_j \leq 0$ is the partial null hypothesis) since they lead to values of ψ^* not smaller than ψ. The observed value of the global test is $\psi = \psi(p_1, p_2) = 7.1985$, and if we account for the dependence between Y_1 and Y_2 (i.e., if we consider the white dots), there are no other points leading to a larger value of ψ^*. If black dots are considered (assuming independence between Y_1 and Y_2), there are 16 points leading to a value of ψ^* at least equal to ψ. The global p-value is defined as

$$p^G = \frac{1}{C} \sum_{b=1}^C I(\psi^* \geq \psi),$$

where C is the cardinality of the domain of ψ^*. In both cases (white and black dots), Fisher's combining function leads to $p^G = 1/16$. The other curves in Figure 1.3 represent the border of the critical regions when Liptak's (black continuous line) and Tippett's (black dotted line) combining functions are applied. The related global p-values are the ratios between the number of points below each critical curve and the total number of points considered. They are listed in Table 1.1.

Table 1.1. Observed values of ψ and related p-values when ψ^* is defined on $\mathcal{T}$ and on $\mathcal{T}_\perp$.

ψ^*:	Fisher	Liptak	Tippett
ψ	7.19853	1.691431	0.9375
$p^G\|\mathcal{T}$	1/16	1/16	1/8
$p^G\|\mathcal{T}_\perp$	1/16	19/256	31/256

1.4.1 Properties of the Nonparametric Combination Tests

The multivariate permutation tests maintain the properties of univariate permutation tests. The exactness of the global test is guaranteed from the fact that each point in the permutation sample space $\mathcal{Y}/\mathbf{y}$ (which here is q-dimensional) is equally likely under the global null hypothesis. This fact implies that

$$\alpha = \Pr[\psi^* \geq \psi | \mathbf{Y}(\mathbf{0})],$$

where $\mathbf{Y}(\mathbf{0})$ denotes data under the global null hypothesis.

As regards unbiasedness, let T_j be a partial test to assess H_{0j} against H_{1j}, and let p_j be the related p-value, $j = 1,\ldots,q$. Recall from Section 1.3 that, if T_j is unbiased, then

$$F_{p_j|1}(u) = \Pr[p_j \leq u | \mathbf{y}_j(\delta_j)] \geq \Pr[p_j \leq u | \mathbf{y}_j(0)] = F_{p_{j|0}}(u);$$

that is, the p-value distribution under the alternative H_{1j} is stochastically smaller than under the null hypothesis H_{0j}. Therefore, if H_{1j} is true, then $p_{j|1} \overset{d}{\leq} p_{j|0}$ and, ceteris paribus,

$$\psi_1 = \psi(\ldots,p_{j|1},\ldots) \overset{d}{\geq} \psi(\ldots,p_{j|0},\ldots) = \psi_0$$

from property (2) of the combining function ψ. This implies that

$$\Pr(\psi^* \geq \psi | \mathbf{Y}(\delta)) \geq \Pr(\psi^* \geq \psi | \mathbf{Y}(\mathbf{0})) = \alpha.$$

As regards consistency, if partial tests T_j are consistent, then

$$\lim_{n_j \to +\infty} \Pr\{p_j - 0 < \epsilon | \mathbf{y}_j(\delta_j)\} = 1 \qquad \forall\, \epsilon > 0.$$

By property (3) of the combining function ψ, if $p_j \overset{p}{\to} 0$,

$$\lim_{n_j \to +\infty} \Pr\{|\psi(\ldots,p_j,\ldots) - \bar{\psi}(\ldots,p_j,\ldots)| < \epsilon | \mathbf{y}_j(\delta_j)\} = 1 \qquad \forall\, \epsilon > 0,$$

where $\bar{\psi}(\ldots,p_j,\ldots)$ is the supremum of the combining function ψ. We are implicitly assuming that the number of Monte Carlo iterations goes to infinity (this allows us to state that $p_j \overset{p}{\to} 0$).

1.5 Examples

In this section, we will see in detail some testing procedures described in this chapter with the R language.

1.5.1 Univariate Permutation Tests

Let's consider a two-sample univariate test. Suppose that a continuous random variable Y has been observed on two samples of sizes $n_1 = 3$ and $n_2 = 4$ from two probability distributions P_1 and P_2 that have different location parameters. We wish to test the hypothesis $P_1 \overset{d}{=} P_2$ under the alternative $P_1 \overset{d}{\neq} P_2$. For instance, we could considere a two-sample test when $P_1 \sim N(0,1)$ and $P_2 \sim N(1,1)$, so H_0 is false:

```
> set.seed(1)
> n1<-3
> n2<-4
> n<-n1+n2
> y1<-rnorm(n1)
> y2<-rnorm(n2,mean=1)
> y<-c(y1,y2)
> label<-rep(c(1,2),c(n1,n2))
> y1
[1] -0.6264538  0.1836433 -0.8356286
> y2
[1] 2.5952808 1.3295078 0.1795316 1.4874291
> label
[1] 1 1 1 2 2 2 2
```

The vector $\mathbf{y}$ is the vector of the observed data $\mathbf{y} = [\mathbf{y_1}, \mathbf{y_2}]$, whereas `label` is the vector of the sample labels. y denotes the orbit $\mathcal{Y}^n/_{\mathbf{y}}$, whose elements have the same probability under H_0. In particular, the probability of a given point $\mathbf{y}^*$ is $1/n! = 0.000198$. Now let $T^* = \bar{y}_1^* - \bar{y}_2^*$ be the test statistic. Clearly T^* induces a partition on $\mathcal{Y}^n/_{\mathbf{y}}$ since two points $\mathbf{y}^* = [\mathbf{y_1^*}, \mathbf{y_2^*}]$ and $\mathbf{y}' = [\mathbf{y_1'}, \mathbf{y_2'}]$ in $\mathcal{Y}^n/_{\mathbf{y}}$ give the same value of T^* if the event $\{\mathbf{y_1^*} = \mathbf{y_2^*}\} \cap \{\mathbf{y_1'} = \mathbf{y_2'}\}$ is observed. This happens when $\mathbf{y_1'}$ and $\mathbf{y_2'}$ are random permutations of $\mathbf{y_1^*}$ and $\mathbf{y_2^*}$, respectively. Therefore, the cardinality of $\mathcal{T}$ is

```
> C<-choose(n,n1)
```

We are now going to explore in detail the permutation reference set $\mathcal{T}$. We can do that either by specifying all the elements in $\mathcal{Y}^n/_{\mathbf{y}}$ or by enumerating the elements of the partition of $\mathcal{Y}^n/_{\mathbf{y}}$ induced by T^*. To do this, we require the library `combinat`.

```
> library(combinat)
> index<-seq(1,n)
> index
[1] 1 2 3 4 5 6 7
> pi1<-matrix(unlist(combn(n,n1)),nrow=C,byrow=TRUE)
> pi2<-t(apply(pi1,1,function(x){index[-x]}))
> pi<-cbind(pi1,pi2)
> pi
       [,1] [,2] [,3] [,4] [,5] [,6] [,7]
  [1,]    1    2    3    4    5    6    7
  [2,]    1    2    4    3    5    6    7
  [3,]    1    2    5    3    4    6    7
  ........................................
 [33,]    4    5    7    1    2    3    6
 [34,]    4    6    7    1    2    3    5
 [35,]    5    6    7    1    2    3    4
```

Here pi1 is a matrix whose rows are all possible ways to combine the elements of index into groups of size $n_1 = 3$, pi2 is the matrix whose rows are the remaining elements of index, and pi is the matrix whose rows are all distinct arrangements of the elements of index into groups of size n_1 and n_2. The set $\mathcal{T}$ is given by the image $T^*(\pi(\mathbf{y}))$ through the statistic T^*, where $\pi(\cdot)$ is the index permutation in pi:

```
> y.perm<-matrix(y[pi],nrow=C,byrow=F)
> round(y.perm,digits=5)

          [,1]      [,2]      [,3]      [,4]      [,5]      [,6]      [,7]
 [1,] -0.62645   0.18364  -0.83563   2.59528   1.32951   0.17953   1.48743
 [2,] -0.62645   0.18364   2.59528  -0.83563   1.32951   0.17953   1.48743
 [3,] -0.62645   0.18364   1.32951  -0.83563   2.59528   0.17953   1.48743
 ......................................................................
[33,]  2.59528   1.32951   1.48743  -0.62645   0.18364  -0.83563   0.17953
[34,]  2.59528   0.17953   1.48743  -0.62645   0.18364  -0.83563   1.32951
[35,]  1.32951   0.17953   1.48743  -0.62645   0.18364  -0.83563   2.59528
```

Now define the statistic $T^* = \bar{y}_1^* - \bar{y}_2^*$:

```
> T<-apply(y.perm,1,function(x){mean(x[c(1:n1)])
+ -mean(x[-c(1:n1)])})
> T<-array(T,dim=c(C,1))
> T
 [1,] -1.824083677
 [2,]  0.177280148
 [3,] -0.561087453
 ..........
[33,]  2.078799413
[34,]  1.407979989
```

[35,] 0.669612388

Note that the first element of $\mathcal{T}$ (which is T[1,1] in the R environment) is the observed value of the test statistic $T(\mathbf{y}) = \bar{y}_1 - \bar{y}_2$. Now let

```
> sigma0<-sqrt(sum(y^2)/n-mean(y)^2)
```

be the conditional standard deviation of y_{i1}, $i = 1, \ldots, n_1$. The theoretical conditional mean and vairance of T are 0 and $s_0^2(n_1^{-1} + n_2^{-1})$, respectively, where $s_0^2 = n\sigma_0^2/(n-1)$. In fact (leaving out the approximations well known by the R users)

```
> sigma0^2*n*(1/n1+1/n2)/(n-1)
[1] 0.8958433
> mean(T)
[1] 3.220080e-18
> sum(T^2)/C-mean(T)^2
[1] 0.8958433
```

A nonparametric estimate of the probability distribution of T is shown in Figure 1.4. This figure can be obtained by typing plot(density(T,bw=1)), where the argument bw of the density function is the bandwidth of the kernel estimator.

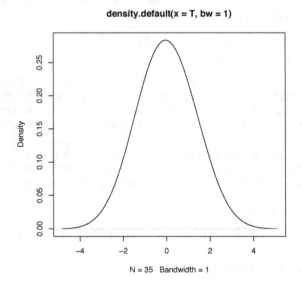

Fig. 1.4. Kernel estimation of the density of T.

In order to obtain a p-value for $H_0 : \delta = 0$ vs. $H_1 : \delta \neq 0$, we may consider the test statistic $T^\star = T^{\star 2}$, which is significant for large values (whereas $T^\star$ is significant for large positive values), and obtain the p-value accordingly:

```
> T.star<-T^2
> p<-(sum(T.star >= T.star[1,]))/C
> p
[1] 0.08571429
```

Of course, if the alternative hypothesis is one-sided, the p-value should be obtained accordingly (i.e., if $H_1 : \delta > 0$, so that small values of T are significant), then

```
> p<-sum(T<=T[1,])/C
> p
[1] 0.05714286
```

A Monte Carlo algorithm to repeat the same test by considering $B = 1000$ MC permutations is given in what follows:

```
>    set.seed(11)
>    B<-1000
>    T<-array(0,dim=c((B+1),1))
>    T[1,]<-mean(y[1:n1])-mean(y[(n1+1):n])
>    for(bb in 2:(B+1)){
+    y.perm<-sample(y)
+    T[bb,]<-mean(y.perm[1:n1])-mean(y.perm[(n1+1):n])
+    }
>    T.star<-T^2
>    p<-sum(T.star[-1,]>=T.star[1,])/B
>    p
[1] 0.085
```

It is worth noting that the p-value obtained by the Monte Carlo algorithm is an estimate of the true p-value, which was obtained previously. Moreover, since the support of T^* is finite, so is the cardinality of $\mathcal{T}$. Therefore, the attainable (exact) p-values are multiples of $1/C$, where $C = \#(\mathcal{T})$ (in our example, $p = 3/35$). Note that the null hypothesis would be rejected at an α-level of 5% only when the value of the observed statistic is the largest one of T.star. Therefore, one has to make conclusions carefully when dealing with a permutation test, especially when n is small.

In the MC algorithm, the vector T has $B + 1$ elements, where the first one corresponds to the observed value of the test statistic, and the remaining ones are obtained from random permutations of **y**. Note that the observed value $T(\mathbf{y})$ is *always* included in T, whereas this is not always the case when the MC algorithm is applied (especially when n is large). That is why the observed value is generally excluded from the null distribution (through the command T.star[-1,]). Another choice is to include the observed value and to compute the p-vaues as p<-sum(T.star>=T.star[1,])/(B+1). Obviously, the greater n, the smaller the difference among the p-values obtained with different methods.

1.5.2 The Nonparametric Combination Methodology

This section will show in detail the bivariate example of Section 1.4. Let $q = 2$, and suppose that we wish to assess the global null hypothesis:

$$H_0^G = \begin{cases} H_{01} : \mu_1 \leq 0 \\ H_{02} : \mu_2 \leq 0 \end{cases} .$$

We will generate data from a bivariate normal distribution with $\mu = [0, 1]'$, although this is not strictly necessary. We also set the correlation ρ_{12} between Y_1 and Y_2 equal to 0.5.

```
> library(combinat)
> library(mvtnorm)
> n<-4
> q<-2
> rho <- 0.5
> I<-diag(q)
> J<-array(1,dim=c(q,q))
> S<-rho*J+(1-rho)*I
> S
     [,1] [,2]
[1,]  1.0  0.5
[2,]  0.5  1.0
> mu = c(0,1)
> set.seed(101)
> y<-rmvnorm(n,mean=mu,sigma=S)
> y
          [,1]        [,2]
[1,] -0.2344941 1.2157956
[2,]  0.8374820 2.2769520
[3,] -0.4917911 1.4230168
[4,]  0.1778775 0.9465873
```

Now let's obtain, for each variable, the permutation distrbution of the test statistic (1.3). To do this, we must obtain all possible combinations of 4 $\pm$ signs. This can be done through the library combinat:

```
> C<-2^n
> z<-rep(2,n)
> sgn<- hcube(z)%/%2
> sgn<-apply(sgn,2,function(x){ifelse(x==0,1,-1)})
> sgn
     [,1] [,2] [,3] [,4]
[1,]    1    1    1    1
[2,]   -1    1    1    1
[3,]    1   -1    1    1
```

```
 [4,]    -1    -1     1     1
 [5,]     1     1    -1     1
 [6,]    -1     1    -1     1
 [7,]     1    -1    -1     1
 [8,]    -1    -1    -1     1
 [9,]     1     1     1    -1
[10,]    -1     1     1    -1
[11,]     1    -1     1    -1
[12,]    -1    -1     1    -1
[13,]     1     1    -1    -1
[14,]    -1     1    -1    -1
[15,]     1    -1    -1    -1
[16,]    -1    -1    -1    -1
```

The null distribution of the test statistic (1.3) can be obtained by multiplying the centered data $y_{ij} - \mu_{0j}$ for the signs in the rows of sgn, and then by adding up together these data. This means that the joint permutation distribution of T_1^* and T_2^* can be directly obtained from the matrix product:

```
>   T.star<-array(0,dim=c(C,2))
>   T.star[,1]<-sgn%*%y[,1]
>   T.star[,2]<-sgn%*%y[,2]
>   T.star
              [,1]        [,2]
 [1,]   0.28907439   5.862352
 [2,]   0.75806254   3.430761
 [3,]  -1.38588962   1.308448
 [4,]  -0.91690147  -1.123143
 [5,]   1.27265657   3.016318
 [6,]   1.74164472   0.584727
 [7,]  -0.40230744  -1.537586
 [8,]   0.06668071  -3.969177
 [9,]  -0.06668071   3.969177
[10,]   0.40230744   1.537586
[11,]  -1.74164472  -0.584727
[12,]  -1.27265657  -3.016318
[13,]   0.91690147   1.123143
[14,]   1.38588962  -1.308448
[15,]  -0.75806254  -3.430761
[16,]  -0.28907439  -5.862352
```

As regards the mean and variance of T_1^* and T_2^*:

```
>   apply(T.star,2,mean)
[1] 0 0
>   apply(T.star,2,var)*(C-1)/C
[1] 1.029862 9.583674
```

```
> apply(y,2,function(x){sum(x^2)})
[1] 1.029862 9.583674
```

Note that the variance of T_j^* depends also on the testing target value μ_{0j}. The points of the space $\mathcal{T}_\perp$ are given by the Cartesian product of T.star[,1] and T.star[,2], which can be obtained by the R function combinat. Instead, the coordinates of the points in the space $\mathcal{T}$ are given by the rows of T.star itself (since the same sign permutations apply). Figure 1.1 has been obtained as follows:

```
> psi<-apply(T.star,1,sum)
> T.ind<-expand.grid(T.star[,1],T.star[,2])
> plot(T.ind,pch=16,xlim=c(-3,3),xlab="T*[,1]",ylab="T*[,2]")
> points(T.star,col="white",pch=16)
> points(T.star,col="black")
> text(T.star[1,1],5.3,"T*[1,]")
```

Now, the observed value of the global test statistic $\psi = T_1(\mathbf{y_1},0) + T_1(\mathbf{y_2},0)$ is

```
> psi[1]
[1] 6.151426
```

and any point whose coordinates (t_x, t_y) satisfy the equation $t_x + t_y = \psi$ has the same significance of ψ against H_0^G. We may thus obtain the line $t_y = \psi - t_x$ for a given grid of 100 points and add it to the plot:

```
> tx<-seq(-3,3,length.out=100)
> lines(tx,psi[1]-tx,type="l",col="grey")
> for(i in 1:100){
+ lines(c(tx[i],tx[i]),c(psi[1]-tx[i],10),type="l",col="grey")
+ }
```

The p-values related to the partial null hypotheses are obtained as follows:

```
> partial.p<-(C+1-apply(T.star,2,rank))/C
> partial.p
         [,1]    [,2]
 [1,]  0.4375  0.0625
 [2,]  0.3125  0.1875
 [3,]  0.9375  0.3750
 [4,]  0.8125  0.6250
 [5,]  0.1875  0.2500
 [6,]  0.0625  0.5000
 [7,]  0.6875  0.7500
 [8,]  0.5000  0.9375
 [9,]  0.5625  0.1250
[10,]  0.3750  0.3125
[11,]  1.0000  0.5625
```

```
[12,] 0.8750 0.8125
[13,] 0.2500 0.4375
[14,] 0.1250 0.6875
[15,] 0.7500 0.8750
[16,] 0.6250 1.0000
```

The element of the ith row and jth column of `partial.p` is obtained as follows:

$$p_{ij}^* = \frac{1}{C} \sum_{l=1}^{C} I(T_{lj}^* \geq T_{ij}^*) \qquad j = 1, 2; i = 1, \ldots, C.$$

The first row of `partial.p` is the vector of observed partial p-values. Thus, there is no evidence against H_{01}, whereas there is evidence against H_{02} (be reminded that the minimum achievable p-value in the permutation framework is $1/C$). The coordinates of the white points in Figure 1.3 are exactly the rows of `partial.p`. The dependence between Y_1 and Y_2 is still maintained because `partial.p` has been obtained by applying the same sign permutations to $\mathbf{y_1}$ and $\mathbf{y_2}$. The coordinates of the black points are given by the Cartesian product of the columns of `partial.p`. To draw Figure 1.3, the following instructions are to be typed:

```
> P.ind<-expand.grid(partial.p[,1],partial.p[,2])
> plot(P.ind,pch=16,xlab="p*[,1]",ylab="p*[,2]",xlim=c(0,1),
+ ylim=c(0,1))
> points(partial.p,col="white",pch=16)
> points(partial.p,col="black")
> text(partial.p[1,1], 0.1,"p[1,]")
```

1.6 Multiple Testing

Multiple testing refers to the testing of more than one hypothesis at a time. The whole subject of multiple hypothesis testing is frequently, and somewhat inaccurately, called either "multiple comparisons" or "multiple tests". Here, by "multiple comparisons" we will be referring to comparisons among different groups, whereas by "multiple tests" we mean the whole subject of multiple testing, but often in the context of multivariate data.

Consider the general problem of simultaneously testing a finite number of hypotheses $H_j, j = 1, \ldots, k$. A classical approach requires that the probability of rejecting one or more true null hypotheses not exceed a given level α (Hochberg and Tamhane, 1987). This probability is called the *familywise error rate* (FWE). Here the term "family" refers to the collection of hypotheses $\{H_1, \ldots, H_k\}$ that is being considered for joint testing. Let $K \subseteq \{1, \ldots, k\}$, and suppose hypotheses H_j with $j \in K \subseteq \{1, \ldots, k\}$ are true and the remainder false. We shall require that

$$\text{FWE} = \Pr\left(\text{reject any } H_j : j \in K\right) \leq \alpha. \tag{1.6}$$

This constraint must hold for all possible configurations of true and null hypotheses; that is, we demand *strong control* of the FWE to distinguish it from the weaker condition of *weak control*, which requires (1.6) to hold only when all the hypotheses of the family are true.

The Bonferroni adjustment is the most basic procedure that controls the FWE. Effectively, the Bonferroni procedure consists of multiplying each individual p-value by k. The Bonferroni method is an example of a *single-step* procedure, meaning that any hypothesis is rejected if its corresponding p-value is less than a common cutoff value (which in the Bonferroni case is α/k). Holm (1979) improved this single-step procedure by the following *step-down* procedure, which we now briefly describe: Order the p-values as $p_{(1)} \leq \cdots \leq p_{(k)}$, and let $H_{(1)}, \ldots, H_{(k)}$ denote the corresponding hypotheses. Then $H_{(1)}, \ldots, H_{(r)}$ are rejected if $p_{(j)} \leq \alpha/(k - j + 1)$ for $j = 1, \ldots, r$, and the remainder are accepted if $p_{(r+1)} > \alpha/(k - r)$. However, one can use Holm's procedure to control the FWER conservatively because it holds under any joint distribution of the p-values, including the worst possible. To improve on Holm's method, Westfall and Young (1993) made effective use of resampling procedures to estimate the joint distributions of multiple test statistics, but under the assumption of *subset pivotality*. Whatever the underlying assumption, by using a bootstrap-based procedure, we achieve control of the FWE only in an asymptotic sense. We focus instead on a permutation-based procedure (Romano and Wolf, 2005), that provides exact FWE control.

The closure method of Marcus et al. (1976) provides a general strategy for constructing a valid multiple test procedure that controls the FWE. Suppose that test statistics T_j for H_j are available, and for any $K \subseteq \{1, \ldots k\}$, let $H_K := \bigcap_{j \in K} H_j$ denote the hypothesis that all H_j with $j \in K$ are true and T_K denote a test statistic for H_K, which can be a function of test statistics or p-values, for which large values of T_K indicate evidence against H_K. The idea behind closed testing is simple: You may reject any hypothesis H_j, while controlling the FWE, when the test of H_j itself is significant and the test of every intersection hypothesis that includes H_j is significant. The closure method using permutation tests works as follows:

$$
\begin{array}{ccc}
 & H_{123} & \\
\swarrow & \downarrow & \searrow \\
H_{12} & H_{13} & H_{23} \\
\downarrow \times & & \times \downarrow \\
H_1 & H_2 & H_3
\end{array}
$$

Suppose you want to test the family $\{H_1, H_2, H_3\}$, where $H_j : Y_{1j} \overset{d}{=} Y_{2j}$, for instance, representing a comparison of a treatment with a control using three distinct measurements. The diagram above shows the closure of the family arranged in a hierarchical fashion, to better illustrate the closed testing method.

(Closed testing)

- Create the closure of the set, which is the set of all possible intersection hypotheses.
- Test all the hypotheses simultaneously by using permutation tests:
    1. Compute the statistics T_K for each nonempty $K \subseteq \{1, \ldots, k\}$.
    2. For b from 1 to B,
        a) perform the bth permutation of treatment and control labels and
        b) compute the statistics $T_K^*(b)$ for each nonempty $K \subseteq \{1, \ldots, k\}$ on the bth permutation of the data.
    3. Compute the "raw" p-values as

$$p_K = \frac{\#\{T_K^*(b) \geq T_K\}}{B}.$$

- Reject any hypothesis H_j, with control of the FWE, when the test of H_j itself is significant and the test of every intersection hypothesis that includes H_j is significant.

Consider the following data set, taken from Westfall et al. (1999):

Table 1.2. Westfall et al. data.

x	Y_1	Y_2	Y_3
1	14.4	7.00	4.30
1	14.6	7.09	3.88
1	13.8	7.06	5.34
1	10.1	4.26	4.26
1	11.1	5.49	4.52
1	12.4	6.13	5.69
1	12.7	6.69	4.45
2	11.8	5.44	3.94
2	18.3	1.28	0.67
2	18.0	1.50	0.67
2	20.8	1.51	0.72
2	18.3	1.14	0.67
2	14.8	2.74	0.67
2	13.8	7.08	3.43
2	11.5	6.37	5.64
2	10.9	6.26	3.47

Let $T_K = \sum_{j \in K} T_j^2$ be the statistic testing $H_K : \bigcap_{j \in K} \left\{ Y_{1j} \stackrel{\mathrm{d}}{=} Y_{2j} \right\}$ against

the two-sided alternative $H_K' : \bigcup_{j \in K} \left\{ Y_{1j} \stackrel{\mathrm{d}}{\neq} Y_{2j} \right\}$, where

$$T_j = \frac{\bar{Y}_{1j} - \bar{Y}_{2j}}{\hat{\sigma}_j \sqrt{\frac{1}{n_1} + \frac{1}{n_2}}}$$

is Student's t statistic comparing the control with the treatment for the jth component variable. An informative way of reporting the results of a closed testing procedure is given by the adjusted p-value.

Definition 1.1 (Adjusted p-value). *The adjusted p-value for a given hypothesis H_j is the maximum of all raw p-values of tests that include H_j as a special case (including the raw p-value of the H_j test itself).*

The following diagram also shows the raw p-values for the hypotheses. The squared areas show how to compute the adjusted p-value for H_3. You must obtain a statistically significant result for the H_3 test itself, as well as a significant result for all hypotheses that include H_3: In this case, H_{13}, H_{23}, and H_{123}. The adjusted p-value for testing H_3 is therefore formally computed as $\max(0.0094, 0.0194, 0.0144, 0.0198) = 0.0198$. Similar reasoning shows that the adjusted p-value for H_2 is 0.0458 and that of H_1 is 0.0966, so that H_2 and H_3 may be rejected at the $\alpha = 0.05$ level while controlling the FWE.

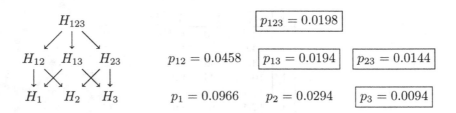

By the closure method of Marcus et al. (1976), the problem shifts to appropriately testing each intersection hypothesis H_K; that is, constructing a single test that controls the usual probability of type I error. Then there is not only the computational issue of constructing a large number of tests, but also the question of an appropriate resampling mechanism that obeys the null hypothesis.

It is often useful to find shortcut procedures that can reduce the number of operations to the order of k. Lehmann and Romano (2005) considered a generic step-down procedure:

(Generic Step-down Procedure)

0. Let $K_0 = \{1, \ldots, k\}$.

   If $T_j \leq c_{K_0}(\alpha)$ for all $j \in K_0$, then accept all hypotheses and stop; otherwise reject each H_j for which $T_i > c_{K_0}(\alpha)$ and continue.

s. Let K_s be the indices of hypotheses not rejected in step $s - 1$.

   If $T_j \leq c_{K_s}(\alpha)$ for all $j \in K_s$, then accept all remaining hypotheses and stop; otherwise reject each H_j for which $T_j > c_{K_s}(\alpha)$ and continue at step $s + 1$.

To control the FWE strongly, the procedure requires two conditions (Lehmann and Romano, 2005, Theorem 9.1.3), given in Theorem 1.2.

Theorem 1.2. *Consider the step-down procedure described in Algorithm 2. The following two conditions are sufficient for controlling the FWE at α:*

(i) monotonicity of the critical values:

$$\text{for every } K \subset K', \quad c_K(\alpha) \leq c_{K'}(\alpha);$$

(ii) weak control of the familywise error at each step:

$$\text{when } H_K \text{ is true,} \quad \Pr(\max(T_j : j \in K) > c_K(\alpha)) \leq \alpha.$$

For a nominal level α, the critical value c_K is defined as the smallest qth value among the permutation distributions of $T_K = \max(T_j : j \in K)$,

$$c_K(\alpha) = \{\max(T_j^*(b) : j \in K), b = 1, \ldots, B\}_{(q)},$$

with $q = B - \lfloor B\alpha \rfloor$, where $\lfloor \cdot \rfloor$ is the largest integer that is at most equal to $B\alpha$. Consequently, the critical values defined in (1.7) satisfy the monotonicity requirement (i) of Theorem 1.2.

In order to satisfy also requirement (ii) of Theorem 1.2, the model considered must guarantee the *randomization hypothesis* (Lehmann and Romano, 2005, Definition 15.2.1). The randomization hypothesis says that under the null hypothesis H_K, the distribution of the subset of data that is used for the calculation of $(T_j : j \in K)$ is not affected by the transformations considered (here, permutations). Thus, we require *exchangeability* of the observed data. The behavior of permutation tests when the randomization hypothesis does not hold was studied in the univariate and multivariate cases by Romano (1990) and Huang et al. (2006), respectively. We present an example showing the failure of the randomization hypothesis.

Example 1.3 (A two-sample problem with k variables).

Suppose that $Y_{11}, \ldots, Y_{1n_1}$ is a control sample of n_1 independent observations and that $Y_{21}, \ldots, Y_{2n_2}$ is a treatment sample of n_2 observations, with $N = n_1 + n_2$. The problem is to test simultaneously the k hypotheses

$$H_j : Y_{1j} \stackrel{\mathrm{d}}{=} Y_{2j}, \qquad j = 1, \dots, k.$$

Note that H_j pertains only to the marginal distributions Y_{1j} and Y_{2j}; nothing else can be said about the joint distributions $(Y_{1j} : j \in K)$ and $(Y_{2j} : j \in K)$ for any $K \subseteq \{1, \dots k\}$. Indeed, we know that $(Y_{1,j} : j \in K) \stackrel{\mathrm{d}}{=} (Y_{2,j} : j \in K) \Rightarrow H_K : \bigcap_{j \in K} \left\{ Y_{1j} \stackrel{\mathrm{d}}{=} Y_{2j} \right\}$, but the reverse implication is not necessarily true. This interpretation matches the setup of the classical multiple testing problem. However, the basis for permutation testing is to assume the narrower null hypothesis that the samples are generated from the same probability law (i.e., $(Y_{1,j} : j \in K) \stackrel{\mathrm{d}}{=} (Y_{2,j} : j \in K)$), and thus the observations can be permuted to either of the two groups and the distribution of the permuted samples is the same as the distribution of the original samples, so that the randomization hypothesis holds.

A location shift model (that is, assuming $\mathbf{Y}_1 \stackrel{\mathrm{d}}{=} \mathbf{Y}_2 - \boldsymbol{\delta}$ with $\boldsymbol{\delta} = (\delta_1, \dots, \delta_k) \in I\!\!R^k$) allows for a permutation-based multiple testing construction because $H_K : \bigcap_{j \in K} \left\{ Y_{1j} \stackrel{\mathrm{d}}{=} Y_{2j} \right\} \Leftrightarrow (Y_{1j} : j \in K) \stackrel{\mathrm{d}}{=} (Y_{2j} : j \in K)$. Now H_j can be viewed as testing $\delta_j = 0$, and the model does not assume knowledge of the form of the underlying distribution.

However, the multivariate normal model $\mathbf{Y}_i \sim N_k(\boldsymbol{\mu}_i, \boldsymbol{\Sigma}_i)$ allows for asymptotically valid multiple testing inference on the difference in means $H_j : \mu_{1j} = \mu_{2j}$ if $n_1 = n_2$, even though under heteroscedasticity $\boldsymbol{\Sigma}_1 \neq \boldsymbol{\Sigma}_2$. Huang et al. (2006) showed that under the null hypothesis of equality of means $\mu_{1j} = \mu_{2j}$, $j = 1, \dots, k$, the vector of test statistic $(T_1, \dots, T_k)$ with $T_j = \bar{Y}_{1j} - \bar{Y}_{2j}$ is distributed as $N_k \left(\mathbf{0}, \frac{\boldsymbol{\Sigma}_1}{n_1} + \frac{\boldsymbol{\Sigma}_2}{n_2} \right)$, whereas the permutation distribution is asymptotically

$$\sum_{r=0}^{n_1} \frac{\binom{n_1}{r}\binom{n_2}{r}}{\binom{N}{n_1}} N_k \left(\mathbf{0}, \frac{(n_1 - r)\boldsymbol{\Sigma}_1 + r\boldsymbol{\Sigma}_2}{n_1^2} + \frac{r\boldsymbol{\Sigma}_1 + (n_2 - r)\boldsymbol{\Sigma}_2}{n_2^2} \right).$$

Suppose that $\mathbf{Y}_1$ and $\mathbf{Y}_2$ are arbitrary multivariate distributions; even for finite but balanced samples, the exact distribution of $(T_1, \dots, T_k)$ under $\bigcap_{j=1}^{k} \{Y_{1j} \stackrel{\mathrm{d}}{=} Y_{2j}\}$ has the same correlation matrix as the distribution of $(T_1, \dots, T_k)$ under $\mathbf{Y}_1 \stackrel{\mathrm{d}}{=} \mathbf{Y}_2$ (Pollard and van der Laan, 2004). Hence, under asymptotic normality for our vector of mean score differences, we obtain asymptotically valid tests with the permutation approach, even if the randomization hypothesis does not hold. For small and balanced samples, strictly speaking neither procedure is valid, although the permutation distribution differs from the true distribution in odd cumulants of order three and higher (see Huang et al., 2006, Corollary 2.3), which may have a negligible effect.

Summing up, unless $n_1 = n_2$ or $\mathbf{Y}_1 \stackrel{\mathrm{d}}{=} \mathbf{Y}_2$ holds, we recommend using a bootstrap approach (Pollard and van der Laan, 2004) since it preserves the correlation structure of the original data and it is asymptotically valid.

On the other hand, when a study is balanced, one may quite safely use the permutation approach.

Applications with R functions

Here we provide assistance in doing the multiple testing procedures illustrated in the previous section using the R language. Load Westfall et al.'s data

```
> load("westdata.Rdata")
> ls()
[1] "Y" "x"
```

where Y is a 16×3 matrix representing a comparison of control units (first seven rows) with treatment units (last nine rows) using three distinct measurements (columns) and x is a vector of integers corresponding to observation (rows) class labels, 1 for the control and 2 for the treatment. Perform the ptest2s function for comparing two independent samples on multivariate data based on Student's t statistics T_j, $j = 1, 2, 3$,

```
> source("ptest2s.R")
> set.seed(0)
> B <- 5000
> T <- ptest2s(Y,x,5000, "Student")
> dim(T)
[1] 5000    3
> T <- T^2
```

obtaining a $B \times 3$ matrix corresponding to the values of (T_1, T_2, T_3) obtained from the observed data (first row) and permuted data (remaining rows). Because the alternative hypotheses are two-sided, we considered as test statistics T_j^2, which are significant for large values. The raw p-values are given by

```
> source("t2p.R")
> P <- t2p(T)
> P[1,]
[1] 0.0966 0.0294 0.0094
```

The adjusted p-values can be obtained by performing a closed testing procedure based on $T_K = \sum_{j \in K} T_j^2$,

```
> source("clostest.R")
> adjP <- clostest(T,combi="sum")
> adjP
[1,] 0.0966 0.0458 0.0198
```

on $T_K = \max(T_j^2 : j \in K)$,

```
> adjP <- clostest(T,combi="max")
> adjP
[1,] 0.0966 0.0502 0.019
```

or on Fisher's combination based on p-values, $T_K = \sum_{j \in K} -2\log(p_j)$

```
> m2lP <- -2*log(P)
> adjP <- clostest(m2lP,combi="sum")
> adjP
[1,] 0.0966 0.0438 0.0238
```

It is often useful to use the step-down procedures based on the maximum test statistic $T_K = \max(T_j^2 : j \in K)$

```
> source("stepdown.R")
> adjP <- stepdown(T)
> adjP
[1,] 0.0966 0.0502 0.019
```

or minimum p-value $T_K = \min(p_j : j \in K)$

```
> mP <- -P
> adjP <- stepdown(mP)
> adjP
[1] 0.0966 0.0532 0.0210
```

because they allow for a shortcut in the number of computations; note, however, that the result based on $T_K = \max(T_j^2 : j \in K)$ does not change by using the closed testing or the step-down procedure, but the latter is faster.

1.7 Multiple Comparisons

In a one-way classification involving different treatments, specific comparisons are often of interest to the researcher. Classical examples include pairwise comparisons (i.e., all treatments with each other) and many-to-one comparisons (i.e., competing treatments with a control). Commonly used multiple comparison procedures are discussed in Hochberg and Tamhane's (1987) book, but nonparametric procedures have not been developed to the same extent as their normal theory counterparts. Because the normality assumption is not always valid, there is a need for distribution-free procedures. In particular, the subject of permutation-based multiple comparison procedures controlling the familywise error rate in the strong sense is still far from fully developed.

In the case of pairwise comparisons, Miller (1981) proposed a procedure based on the permutation distribution of the range of the sample means, but Petrondas and Gabriel (1983) showed that this procedure doesn't control the familywise error rate.

In the case of many-to-one comparisons, Westfall et al. (1999) gave an example with Bernoulli responses where the permutation-based procedure of Westfall and Young (1993) performs badly.

Suppose we have the data displayed in Table 1.3, and we are interested in the comparisons of treatments 1 and 2 with the control (treatment 0), with one-sided alternatives specifying higher and lower probabilities of success, respectively.

Table 1.3. Teresa Neeman's example, from Westfall et al. (1999).

Treatment	# success / # trials	Percent
0	3 / 4	75%
1	1 / 4	25%
2	0 / 2000	0%

Here $Y_j \sim$ Bernoulli(θ_j) and the multiple testing problem is defined by $H_1 : \theta_0 = \theta_1$ against $H_1' : \theta_0 < \theta_1$ and $H_2 : \theta_0 = \theta_2$ against $H_2' : \theta_0 > \theta_2$. The statistic testing H_j is defined as $T_i = -p_i$, where p_i is the p-value from Fisher's one-sided exact test. By performing `PROC MULTTEST` implemented in SAS, the output is as follows:

```
Contrast      Raw_p        StepPerm_p

1 - 0         0.9857       0.0076
0 - 2         0.0001       0.0001
```

For the second comparison, the procedure gives a multiplicity-adjusted p-value of 0.0076, supporting the alternative hypothesis that treatment 1 has a greater success probability than the control. Westfall et al. (1999) commented on this counterintuitive result as a consequence of failure of the subset pivotality condition.

Under the "complete" null $H_{K_0} : \bigcap_{j=1}^{k} Y_0 \overset{d}{=} Y_j$, we can randomly assign all $\sum_{j=0}^{k} n_j$ observations to any of the groups; that is, the group of transformations consisting of all permutations of the data. However, if only a subset of the hypotheses are true, then this group of transformations is not valid. Under $H_K : \bigcap_{i \in K} Y_0 \overset{d}{=} Y_j$, a *valid* group of transformations such that the randomization hypothesis holds consists of those permutations that permute observations within the sample 0 and the samples $j \in \mathcal{K}$ (Romano and Wolf, 2005, Example 6). As a consequence, for $K \subset K'$, the permutation joint distribution of $(T_j : j \in K)$ under H_K is not the same as the permutation joint distribution of $(T_j : j \in K)$ under $H_{K'}$. This means that the joint distribution of the test statistics used for testing the hypotheses H_j, $j \in \mathcal{K}$, is affected by the truth or falsehood of the remaining hypotheses. As a consequence, the subset pivotality condition of Westfall and Young (1993) fails. Indeed, subset pivotality requires that the joint distribution of $(T_j : j \in K)$ under H_K be the same as the joint distribution of $(T_j : j \in K)$ under the complete null H_{K_0}.

To illustrate what is going wrong by using the permutation-based Westfall and Young procedure in the case of multiple comparisons, consider Teresa Neeman's example. The observed test statistics are $T_1 = -0.9857$ and $T_2 =$

-0.0001. For a nominal α level, say 5%, the procedure starts by rejecting the "complete" null $H_{\{1,2\}} : H_1 \cap H_2$ based on the critical value $c_{\{1,2\}}(\alpha) = -1$. This is because there are so many observations in the second group, all zeros, and thus most permutations will have the four occurrences in the second treatment and the p-values for the $1-0$ and $0-2$ comparisons will be 1 since we compare $0/4$ with $0/4$ and $0/4$ with $4/2000$ by Fisher's one-sided exact tests.

In its single-step form, the procedure rejects both H_1 and H_2 since T_1 and T_2 are greater than $c_{\{1,2\}}(\alpha)$, but because subset pivotality fails, it guarantees only weak control and not strong control (see also Romano and Wolf, 2005, Example 1).

In its stepwise form, once H_2 is rejected, the procedure removes it and test H_1 based on a critical value equal to -1's obtained by still considering all possible permutations that satisfy condition (i) but not condition (ii) of Theorem 1.2. Indeed, because only H_1 is assumed to be true, for the permutations other than those shuffling the first treatment and control observations, the randomization hypothesis doesn't hold.

To fix the problem, by also computing $c_1(\alpha) = -0.75$ by using permutations between samples 0 and 1 and $c_2(\alpha) = -1$ by using permutations between samples 0 and 2, we see that condition (i) is not satisfied. Thus a closed testing procedure is required: We reject only H_2 because T_1 is greater than $c_{\{1,2\}}(\alpha)$ but less than $c_1(\alpha)$. The following diagram gives the raw p-values for the hypotheses, so that the adjusted p-values are 0.9857 and 0.0001 for H_1 and H_2, respectively.

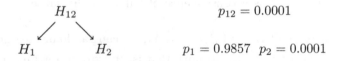

$$H_{12} \qquad\qquad\qquad p_{12} = 0.0001$$

$$H_1 \qquad H_2 \qquad\qquad p_1 = 0.9857 \quad p_2 = 0.0001$$

By testing all the $2^k - 1$ intersection hypotheses $H_{\mathcal{K}}$ by using valid permutations, the closure method provides strong control of the FWE, although this can be computationally prohibitive for large k. As an alternative, one may construct *separately* for each hypothesis H_j the corresponding permutation distribution T_j, obtaining an α-level test for H_j. The collection $\{(H_j, T_j), i = 1, \ldots, k\}$ obtained forms a *testing family* because, for every i, the permutation distribution of T_j is completely specified under H_j. But it is not a *joint testing family* (Hochberg and Tamhane, 1987) because, for every K, the joint distribution of $(T_j : j \in K)$ is not specified under H_K. Indeed, with only the permutation marginal distributions available, nothing is known about the joint distribution of the T_j's. However, one can use Holm's procedure to control the FWE conservatively.

The previous discussion dealt with many-to-one comparisons, so that any subset of the hypotheses $H_1, \ldots, H_k$ can be true, with the remaining ones being false. This condition is not satisfied for all pairwise comparisons $H_{jj'}$:

$Y_j \overset{d}{=} Y_{j'}$ because for instance the set $\{H_{12}, H_{23}\}$ cannot be the set of all true hypotheses since the truth of H_{12} and H_{23} implies the truth of H_{13}. In Holm's terminology, we can say that many-to-one comparisons satisfy the *free combination* condition, whereas all pairwise comparisons represent *restricted combinations*. When the hypotheses are restricted, then certain combinations of true hypotheses necessarily imply truth or falsehood of other hypotheses. In these cases, the adjustments may be made smaller than the free combination adjusted p-values while maintaining strong control of the FWE. For example, Shaffer (1986) modified Holm's procedure to give more powerful tests.

As a final remark, there is a need for better permutation-based procedures, and it is reasonable to explore the stepwise approach incorporating the dependence among test statistics in an attempt to improve the power while maintaining FWE control.

Part I

Stochastic Ordering

2

Ordinal Data

2.1 Introduction

A categorical variable has a measurement scale consisting of a set of categories. Categorical variables that have ordered categories are called *ordinal* (Agresti, 2002). They appear, for example, whenever the condition of a patient cannot be measured by a metric variable and has to be classified or rated as "critical", "serious", "fair", or "good". The measurements on ordered categorical scales can be ordered by size, but the scales lack any algebraic structure; that is, the distances between categories are unknown. Although a patient categorized as "fair" is more healthy than a patient categorized as "serious", no numerical value describes how much more healthy that patient is.

Let X and Y denote two categorical variables, X with r categories and Y with c categories. The classification of n measurements on both variables has rc possible combinations, which can be represented in an $r \times c$ contingency table (see Table 2.1), where $\{m_{i,j}, i = 1, \ldots, r, j = 1, \ldots, c\}$ represents cell frequencies, with row and column margins $n_i = \sum_{j=1}^{c} m_{i,j}$ and $t_j = \sum_{i=1}^{r} m_{i,j}$, respectively.

Table 2.1. $r \times c$ contingency table.

	1	$\cdots$	j	$\cdots$	c	
1	$m_{1,1}$	$\cdots$	$m_{1,j}$	$\cdots$	$m_{1,c}$	n_1
$\vdots$	$\vdots$	$\vdots$	$\vdots$	$\vdots$	$\vdots$	$\vdots$
i	$m_{i,1}$	$\cdots$	$m_{i,j}$	$\cdots$	$m_{i,c}$	n_i
$\vdots$	$\vdots$	$\vdots$	$\vdots$	$\vdots$	$\vdots$	$\vdots$
r	$m_{r,1}$	$\cdots$	$m_{r,j}$	$\cdots$	$m_{r,c}$	n_r
	t_1	$\cdots$	t_j	$\cdots$	t_c	n

D. Basso et al., *Permutation Tests for Stochastic Ordering and ANOVA*, Lecture Notes in Statistics, 194, DOI 10.1007/978-0-387-85956-9_2,
© Springer Science+Business Media, LLC 2009

There is a large body of literature concerning the analysis of categorical data for which the row and column variables are ordinal measurements. In recent years, statisticians increasingly have recognized that many benefits can result from using methods that take into account orderings among categories in contingency tables. One way to utilize ordered categories is to assume inequality constraints on parameters for those categories that describe dependence structure.

In many applications, one would typically expect larger values of Y to be associated with larger values of X. One can describe the positive dependence of the discrete bivariate distribution of (X, Y) using various types of odds ratios, referred to as generalized odds ratios. Three of them, which are the most commonly used in application problems, are defined below. Let $i = 1, \ldots, r-1$, $j = 1, \ldots, c-1$. Then:

1. *Local odds ratios*:

$$\theta_{i,j}^L = \frac{\Pr(Y = j | X = i) \Pr(Y = j + 1 | X = i + 1)}{\Pr(Y = j | X = i + 1) \Pr(Y = j + 1 | X = i)}.$$

2. *Cumulative odds ratios*:

$$\theta_{i,j}^C = \frac{\Pr(Y \le j | X = i) \Pr(Y > j | X = i + 1)}{\Pr(Y \le j | X = i + 1) \Pr(Y > j | X = i)}.$$

3. *Global odds ratios*:

$$\theta_{i,j}^G = \frac{\Pr(Y \le j | X \le i) \Pr(Y > j | X > i)}{\Pr(Y \le j | X > i) \Pr(Y > j | X \le i)}.$$

These definitions show that generalized odds ratios are odds ratios for 2×2 tables obtained from the $r \times c$ table by collapsing adjacent categories if necessary, as displayed in Figure 2.1.

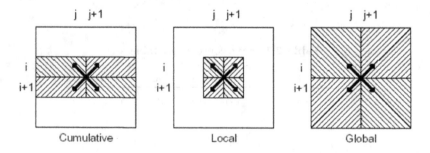

| j j+1 | j j+1 | j j+1 |

Cumulative   Local   Global

Fig. 2.1. Generalized odds ratios.

If $\boldsymbol{\theta}$ denotes the column vector of any one of the generalized odds ratios, then $\boldsymbol{\theta} = \mathbf{1}_{(r-1)(c-1)}$ if and only if X and Y are independent, where $\mathbf{1}_k$ is used to denote a k-dimensional column vector of ones. Relationships among the positive dependencies considered are

$$\theta_1^L \gneq 1_{(r-1)(c-1)} \Rightarrow \theta_1^C \gneq 1_{(r-1)(c-1)} \Rightarrow \theta_1^G \gneq 1_{(r-1)(c-1)},$$

where $\theta \gneq 1_{(r-1)(c-1)}$ means that $\theta_{i,j} \geq 1$, $i = 1, \ldots, r-1$, $j = 1, \ldots, c-1$, with at least one strict inequality.

To introduce the notation to be adopted throughout, suppose that the $r \times c$ table arises from the comparison of r increasing levels of a treatment for which the *response* variable Y is ordinal with c categories. The c levels can be thought of as c ordinal categories of an *explanatory* variable X. A primary aim of many studies is to compare conditional distributions of Y at various levels of the explanatory variable X. For $i = 1, \ldots, r$ and $j = 1, \ldots, c$, let

$$\pi_i(j) = \Pr(Y = j | X = i), \qquad F_i(j) = \sum_{l=1}^{j} \pi_i(l),$$

be the probability mass function (p.m.f.) and cumulative distribution function (c.d.f.) of Y_i, respectively, where Y_i denotes a random variable whose distribution is the conditional distribution of Y given $X = i$.

One of the earliest definitions of stochastic ordering was given by Lehmann (1955).

Definition 2.1. *The random variable Y_{i+1} is said to dominate Y_i according to the simple stochastic ordering, or Y_{i+1} is stochastically larger than Y_i, written $Y_i \overset{st}{\leq} Y_{i+1}$, if*

$$F_i(j) \geq F_{i+1}(j), \qquad j = 1, \ldots, c-1.$$

In some cases, a pair of distributions may satisfy a stronger condition.

Definition 2.2. *The random variable Y_{i+1} is said to dominate Y_i according to the likelihood ratio ordering, written $Y_i \overset{lr}{\leq} Y_{i+1}$, if*

$$\pi_i(j)\pi_{i+1}(j+1) \geq \pi_i(j+1)\pi_{i+1}(j), \qquad j = 1, \ldots, c-1.$$

Clearly, if $\theta^L \geq 1_{(r-1)(c-1)}$ and $\theta^C \geq 1_{(r-1)(c-1)}$, then the rows satisfy the likelihood ratio ordering $Y_1 \overset{lr}{\leq} \ldots \overset{lr}{\leq} Y_r$ and the simple stochastic ordering $Y_1 \overset{st}{\leq} \ldots \overset{st}{\leq} Y_r$, respectively (and vice versa). Each of these constraints defines a form of monotone order of the rows involving only two rows at a time. By contrast, the constraint $\theta^G \geq 1_{(r-1)(c-1)}$ define a monotone relationship that involves more than two rows at a time; this constraint is equivalent to the notion of positive quadrant dependence (PQD, Lehmann, 1955).

Definition 2.3. *We shall say that the pair (X, Y) is positive quadrant dependent if*

$$\Pr(Y \leq j | X \leq i) \geq \Pr(Y \leq j | X > i) \quad \forall i, j.$$

The reason why PQD is a positive dependence concept is that X and Y are more likely to be large together or to be small together compared with X' and Y', where $X \stackrel{d}{=} X'$, $Y \stackrel{d}{=} Y'$, and X' and Y' are independent.

In many applications in which it is believed that certain constraints on the distributions exist, it is reasonable to assume a *stochastic ordering*. The statistical information arising from these constraints, if properly incorporated, makes the statistical inference more efficient than its counterparts, wherein such constraints are ignored.

In this chapter, several data examples are presented to motivate and to provide an overview of the topics.

2.2 Testing Whether Treatment is "Better" than Control: $2 \times c$ Contingency Tables

Patefield (1982) reported the results of a double-blind study concerning the use of Oxprenolol in the treatment of examination stress. Thirty-two students were entered in the study: fifteen were treated with Oxprenolol (treatment) and seventeen were given Diazepam (control). The examination grades were compared with their tutor's prediction; the results are given in Table 2.2.

Table 2.2. Examination results compared with tutor's predictions

		Worse	Same	Better	
		1	2	3	Total
Control	1	6	11	0	17
Treatment	2	2	8	5	15
		8	19	5	32

For this example, one wishes to test the null hypothesis that the treatment and control effects are the same against the one-sided alternative that treatment is in some sense "better" than the placebo. One obstacle to the development of suitable tests is that it is often difficult to be specific as to the notion of "better" (Cohen and Sackrowitz, 2000). In fact, a precise definition of "better" and hence a precise definition of an alternative hypothesis is often not even mentioned in instances where such a testing problem is encountered. In contrast, Cohen et al. (2000) offered various formal definitions of "better" in order to improve understanding of the alternative hypothesis.

Among the formally defined notions for a $2 \times c$ table, the less stringent is the simple stochastic order. Let the testing problem be

$$H_0 : Y_1 \stackrel{d}{=} Y_2 \Leftrightarrow \theta^C \in \Theta_0 = \{\theta^C : \theta^C = 1_{c-1}\}, \tag{2.1}$$

where "$\overset{d}{=}$" means "equal in distribution", against the one-sided alternative

$$H_1 : Y_1 \overset{st}{\lesssim} Y_2 \Leftrightarrow \boldsymbol{\theta}^C \in \Theta_1 = \{\boldsymbol{\theta}^C : \boldsymbol{\theta}^C \gtrsim \mathbf{1}_{c-1}\}. \tag{2.2}$$

Suppose that a test rejects H_0. Then it does follow that there is sufficient statistical evidence to support the claim that there is a difference between the treatment and the control, but it does not follow that there is statistical evidence to accept that the treatment is better than the control (H_1 is true). However, if we make the prior assumption that

the treatment is at least as good as the control

(that is, either H_0 or H_1 is true), then the rejection of "no difference between the treatment and the control" together with the prior assumption that "the treatment is at least as good as the control" (that is, $Y_1 \overset{st}{\leq} Y_2 \Leftrightarrow \boldsymbol{\theta}^C \in \Theta_0 \cup \Theta_1$) would lead to the conclusion that the treatment is better than the control (Silvapulle and Sen, 2005). In other words, we consider a model specifying the distribution of Y to be stochastically ordered with respect to the value of the explanatory variable X; that is, $\{Y_x, Y_{x'} : Y_x \overset{st}{\leq} Y_{x'}$ if $x < x'\}$.

2.2.1 Conditional Distribution

Let $\boldsymbol{M}_i = (M_{i,1}, \ldots, M_{i,c})$, $i = 1, 2$, be independent random vectors having multinomial distributions with cell probabilities $\boldsymbol{\pi}_i = (\pi_i(1), \ldots, \pi_i(c))$. Under the product multinomial model (that is, given the row totals $\boldsymbol{n} = (n_1, n_2)$), when the null hypothesis H_0 is true, the column total $\boldsymbol{t} = (t_1, \ldots, t_c)$ is a completely sufficient statistic (row and column total in the full multinomial model). Testing is carried out by conditioning on row and column totals. This allows the two models, product multinomial and full multinomial, to be treated simultaneously since the conditional distributions are the same (Cohen and Sackrowitz, 2000).

Here, for convenience of notation, we drop the row index i from $\theta_{i,j}$. The conditional distribution of $(M_{1,1}, \ldots, M_{1,c-1})$ given the row and column totals $(\boldsymbol{n}, \boldsymbol{t})$ is the *multivariate noncentral hypergeometric distribution* with p.m.f.

$$\underset{\boldsymbol{\theta}^L}{\Pr}(M_{1,1} = m_{1,1}, \ldots, M_{1,c-1} = m_{1,c-1} | \boldsymbol{n}, \boldsymbol{t})$$

$$= \frac{\binom{t_1}{m_{1,1}} \cdots \binom{t_j}{n_1 - \sum_{j=1}^{c-1} m_{1,j}} \prod_{j=1}^{c-1} \left(\prod_{l=j}^{c-1} \theta_l^L\right)^{m_{1,j}}}{\sum_{(m_{1,1}, \ldots, m_{1,c-1}) \in \mathcal{M}} \binom{t_1}{m_{1,1}} \cdots \binom{t_j}{n_1 - \sum_{j=1}^{c-1} m_{1,j}} \prod_{j=1}^{c-1} \left(\prod_{l=j}^{c-1} \theta_l^L\right)^{m_{1,j}}},$$

where $m_{1,j} \geq 0$, $j = 1, \ldots, c-1$, and

$$\mathcal{M} = \left\{ (m_{1,1}, \ldots, m_{1,c-1}) : n_1 - \sum_{j=1}^{c-1} m_{1,j} \geq 0; t_j - m_{1,j} \geq 0; \right. \tag{2.3}$$

$$\left. n + \sum_{j=1}^{c-1} m_{1,j} - \sum_{j=1}^{c-1} t_j - n_1 \right\}.$$

Note that the conditional distribution has a simple exponential family form that depends only on the natural parameters $\boldsymbol{\nu} = (\nu_1, \ldots, \nu_{(c-1)})$, where

$$\nu_j = \log \left(\frac{\pi_1(j)\pi_2(c)}{\pi_2(j)\pi_1(c)} \right) = \log \left(\prod_{l=j}^{c-1} \theta_l^L \right).$$

As the conditional distribution depends on $(\boldsymbol{\pi}_1, \boldsymbol{\pi}_2)$ only through $\boldsymbol{\nu}$, the conditional hypotheses $H_0^{|(n,t)}$ and $H_1^{|(n,t)}$ must be formulated in terms of $\boldsymbol{\nu}$. We have $H_0 : \boldsymbol{\theta}^C = \mathbf{1}_{c-1} \Leftrightarrow H_0^{|(n,t)} : \boldsymbol{\nu} = \mathbf{0}_{c-1}$, but $H_1 : \boldsymbol{\theta}^C \gneq \mathbf{1}_{c-1} \Rightarrow H_1^{|(n,t)} : \nu_1 > 0$ (or the first nonzero element of $\boldsymbol{\nu}$ when $c > 3$). In contrast, for the likelihood order alternative, $H_1 : \boldsymbol{\theta}^L \gneq \mathbf{1}_{c-1} \Leftrightarrow H_1^{|n,t} : \boldsymbol{\nu} \gneq \mathbf{0}_{c-1}$.

In Patefield's example, there are 54 table configurations with the same marginal totals as Table 2.2. The conditional probabilities are given by

$$\Pr_{\theta_1^L, \theta_2^L} (M_{1,1} = m_{1,1}, M_{1,2} = m_{1,2} | (17, 15), (8, 19, 5))$$

$$= \frac{\binom{8}{m_{1,1}}\binom{19}{m_{1,2}}\binom{5}{17-m_{1,1}-m_{2,2}}(\theta_1^L \theta_2^L)^{m_{1,1}}(\theta_2^L)^{m_{1,2}}}{\sum_{m_{1,1}=0}^{8} \sum_{m_{1,2}=12}^{17} \binom{8}{m_{1,1}}\binom{19}{m_{1,2}}\binom{5}{17-m_{1,1}-m_{2,2}}(\theta_1^L \theta_2^L)^{m_{1,1}}(\theta_2^L)^{m_{1,2}}},$$

where $(m_{1,1}, m_{1,2}) \in \mathcal{M} = \{(m_{1,1}, m_{1,2}) : m_{1,1} = 0, \ldots, 8; m_{1,1} + m_{1,2} = 12, \ldots, 17\}$.

2.2.2 Linear Test Statistics: Choice of Scores

A popular class of linear test statistics is based on the explicit or implicit assignment of scores to the c categories (Graubard and Korn, 1987).

Gail (1974) called a "value system" any real function on the sample space of a multinomial random variable. Consider a nondecreasing real function of Y_i, $w(\cdot) : \{1, \ldots, c\} \to \mathbb{R} : -\infty < w(1) \leq \ldots \leq w(c) < \infty$. Denote by $w(j) := w_j$ the score attached to the jth category and by $\boldsymbol{w} = (w_1, \ldots, w_c)$ the vector of scores, and observe that $\mathrm{E}[w(Y_i)] = \sum_{j=1}^{c} w_j \pi_i(j)$. Thus, the hypotheses are

$$H_0^w : \mathrm{E}[w(Y_1)] = \mathrm{E}[w(Y_2)] \Leftrightarrow \sum_{j=1}^{c-1} (w_{j+1} - w_j)(F_1(j) - F_2(j)) = 0$$

against

$$H_1^w : \mathrm{E}\left[w(Y_1)\right] < \mathrm{E}\left[w(Y_2)\right] \Leftrightarrow \sum_{j=1}^{c-1}(w_{j+1} - w_j)\left(F_1(j) - F_2(j)\right) > 0,$$

by which simplicity is achieved by reducing the multiparameter inference problem to one that involves only a scalar parameter. While this simplifies the testing problem, it could also be expected to have low power at points away from the chosen (through w) direction in the alternative space. Note that because of our prior assumption $\{F_1(j) \geq F_2(j), j = 1, \ldots, c-1\}$, H_0 in (3.1) implies H_0^w and H_1^w implies H_1 in (3.2), and when $w_1 < \ldots < w_c$, also $H_0^w \Rightarrow H_0$ and $H_1 \Rightarrow H_1^w$.

The class of linear test statistics based on w is given by

$$T_w = \frac{\left(\frac{(n-2)n_1 n_2}{n}\right)^{\frac{1}{2}}\left(\sum_{j=1}^{c}\frac{m_{2,j}w_j}{n_2} - \sum_{j=1}^{c}\frac{m_{1,j}w_j}{n_1}\right)}{\left(\sum_{j=1}^{c}(w_j)^2 t_j - \frac{1}{n_1}(\sum_{j=1}^{c}w_j m_{1,j})^2 - \frac{1}{n_2}(\sum_{j=1}^{c}w_j m_{2,j})^2\right)^{\frac{1}{2}}};$$

that is, the usual two-sample t statistic based on assigning a set of scores to the c categories. It is straightforward to show that, for any linear transformation of scores that preserves the monotonicity, $T_{a1+bw} = T_w$ with $a \in \mathcal{R}, b \in \mathcal{R}^+$. Hence, we may consider standardized scores w obtained by transforming the original scores via $a = -w_1/(w_c - w_1)$ and $b = 1/(w_c - w_1)$ to the $[0, 1]$ interval. Permutationally equivalent formulations of T_w are

- Graubard and Korn (1987): $T_w = \sum_{j=1}^{c} m_{2,j}w_j$,
- goodness of fit statistics: $T_w = \sum_{k=1}^{c-1}(w_{j+1} - w_j)\left(\hat{F}_1(j) - \hat{F}_2(j)\right)$, and
- Mantel (1963): $T_w = (n-1)^{\frac{1}{2}}\hat{\rho}$,

where $\hat{F}_i(j) = (\sum_{l=1}^{j} m_{i,j})/n_i$ denotes the empirical c.d.f. of the ith group and $\hat{\rho}$ the Pearson correlation coefficient based on the scores w and values 0 and 1 assigned to the control and the treatment, respectively.

Widely used scoring systems in data analysis include equal-spacing scores $w = (1, \ldots, c)$, midrank scores $w = (\bar{r}_1, \ldots, \bar{r}_c)$, where $\bar{r}_1 = \frac{t_1+1}{2}$ and $\bar{r}_j = \sum_{l=1}^{j-1} t_l + \frac{t_j+1}{2}$, $j = 2, \ldots, c$, and Anderson-Darling scores $(w_{j+1} - w_j) = 1/(\hat{F}(j)(1 - \hat{F}(j)))^{1/2}$, $j = 1, \ldots, c$, where $\hat{F}(j) = [n_1\hat{F}_1(j) + n_2\hat{F}_2(j)]/n$.

The use of midrank scores seems appealing since it yields to the Wilcoxon-Mann-Whitney (WMW) test statistic. This test statistic can also be viewed as Spearman's correlation coefficient between X and Y. However, midrank scores do not necessarily provide distances between categories that correspond to a "reasonable" metric. In particular, for highly unbalanced response frequencies, adjacent categories having relatively few observations necessarily have similar midrank scores. For example, suppose few subjects fall in the first categories on the scale "bad", "fair", "good", "very good", "excellent"; mid-ranks then have similar scores for categories "bad" and "good".

For Patefield's data, the critical function of the randomized WMW test based on the statistic $T_{(0,27/51,1)}$ is

$$\phi_{T_{(0,27/51,1)}} = \begin{cases} 1 & \bullet \text{ if } 27m_{1,1} + 24(m_{1,1} + m_{2,2}) > 543 \\ 0.53 & \otimes \text{ if } 27m_{1,1} + 24(m_{1,1} + m_{2,2}) = 543 \\ 0 & \circ \text{ if } 27m_{1,1} + 24(m_{1,1} + m_{2,2}) < 543 \end{cases},$$

and the rejection region at a significance level of $\alpha = 0.05$ is given in Figure 2.2 (a). The power of the randomized WMW test ϕ_T given $\boldsymbol{n} = (17, 15), \boldsymbol{t} = (8, 19, 5)$, the *conditional power*, is given by

$$\begin{aligned}
\beta(\theta_1^L, \theta_2^L | (17, 15), (8, 19, 5)) &= E_{\theta_1^L, \theta_2^L} \left[\phi_{T_{(0,27/51,1)}} | (17, 15), (8, 19, 5) \right] \\
&= \Pr_{\theta_1^L, \theta_2^L} (27M_{1,1} + 24(M_{1,1} + M_{1,2}) > 543 | (17, 15), (8, 19, 5)) \\
&\quad + 0.53 \Pr_{\theta_1^L, \theta_2^L} (27M_{1,1} + 24(M_{1,1} + M_{1,2}) = 543 | (17, 15), (8, 19, 5))
\end{aligned}$$

for $(\theta_1^L, \theta_2^L) : \boldsymbol{\nu} = (\log(\theta_1^L \theta_2^L), \log(\theta_2^L))^t \in \Theta_0^{|(\boldsymbol{n},\boldsymbol{t})} \cup \Theta_1^{|(\boldsymbol{n},\boldsymbol{t})}$, where the conditional null and alternative parameter spaces are $\Theta_0^{|(\boldsymbol{n},\boldsymbol{t})} : \{\boldsymbol{\nu} : \log(\theta_1^L \theta_2^L) = 0, \log(\theta_2^L) = 0\}$ and $\Theta_1^{|(\boldsymbol{n},\boldsymbol{t})} : \{\boldsymbol{\nu} : \log(\theta_1^L \theta_2^L) > 0\}$, respectively. The conditional power is depicted in Figure 2.2 (b) as a function of $(\log(\theta_1^L), \log(\theta_2^L))$.

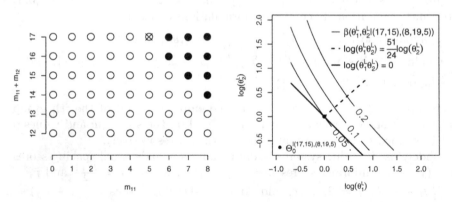

(a) 0.047 level rejection region (nonran-　(b) Contour plot of the conditional power
domized test)　　　　　　　　　　　　　function for the WMW test

Fig. 2.2. WMW test.

Observe that:

- The WMW test is conditionally biased; that is, $\beta(\theta_1^L, \theta_2^L | (17, 15), (8, 19, 5)) < \alpha = 0.05$ for some $(\theta_1^L, \theta_2^L) : \boldsymbol{\nu} \in \Theta_1^{|(\boldsymbol{n},\boldsymbol{t})}$ (see also Berger and Ivanova, 2002b). Note that for the likelihood ratio conditional alternative $H_1^{|\boldsymbol{n},\boldsymbol{t}} : \boldsymbol{\nu} = (\log(\theta_1^L \theta_2^L), \log(\theta_2^L))^t \gneq (0,0)^t$, we have $\Theta_0^{|(\boldsymbol{n},\boldsymbol{t})} \cup \Theta_1^{|(\boldsymbol{n},\boldsymbol{t})} = \{\boldsymbol{\nu} :$

$\log(\theta_1) \geq 0, \log(\theta_2) \geq 0\}$. Cohen and Sackrowitz (1991) showed that the WMW test (among others) is conditionally (and hence unconditionally) unbiased.

- The WMW test is conditionally Bayes with respect to a prior putting all its mass on the set $\{\boldsymbol{\nu} \in \Theta_1^{|(\boldsymbol{n},\boldsymbol{t})} : \log(\theta_1^L \theta_2^L) = \frac{51}{24}\log(\theta_2^L)\}$; that is, the WMW test is very powerful for alternatives near this direction (Cohen and Sackrowitz, 1998, see).

Often it is unclear how to assign scores because the power of the test depends on them. Indeed, the permutation tests listed in Chapter 7 of the StatXact-8 User Manual can be with general scores or with MERT scores (see Section 7.13). Podgor et al. (1996) consider a robust test from several test statistics $T_{\boldsymbol{w}}$ based on different sets of scores. The maximin efficient robust test (MERT) has the property of maximin efficiency in that its lowest asymptotic efficiency relative to each of the possible tests is higher than the lowest such efficiency for any other statistic based on any set of scores. The MERT considers a linear combination of the pair of test statistics with minimum correlation. However, Podgor's MERT is itself a linear rank test, which is ironic since it was proposed to correct the weaknesses of the class of linear rank tests (Berger and Ivanova, 2002a).

To handle the ambiguities arising from the choice of scoring, Kimeldorf et al. (1992) obtained the minimum and the maximum of the $T_{\boldsymbol{w}}$ test statistic over all possible assignments of nondegenerate nondecreasing scores $\boldsymbol{w}$. If the range of min and max values does not include the critical value of the test statistic (they term this case "nonstraddling"), then it can be immediately concluded that the results of the analysis remain the same no matter the choice of increasing scores used. However, if the range includes the critical value (termed the "straddling" case), the choice of scores used in the analysis must be carefully justified.

The scores $\boldsymbol{w}^{\max}$ that maximize $T_{\boldsymbol{w}}$ can be found by considering two cases:

- If $\hat{F}_2 \geq \hat{F}_1$, $\boldsymbol{w}^{\max}$ is one of the $c - 1$ *monotone extreme points*

$$\begin{cases} w_l^{\max} = 0, \, 1 \leq l \leq j \\ w_l^{\max} = 1, \, j+1 \leq l \leq c \end{cases} \quad j = 1, \ldots, c-1.$$

- Otherwise, $w_j^{\max}$, $j = 1, \ldots, c-1$, are given by the *isotonic regression* of $m_{2,j}/t_j$ with weights t_j, denoted by $P_t\left(\frac{\boldsymbol{m_2}}{\boldsymbol{t}}|\mathcal{I}\right)$, the solution that minimizes the weighted sum of squares

$$\min_{\boldsymbol{w} \in \mathcal{I}} \sum_{j=1}^{c} \left(\frac{m_{2,j}}{t_j} - w_j\right)^2 t_j;$$

that is, the weighted least squares projection of $\boldsymbol{m_2}/\boldsymbol{t}$ onto the closed convex cone $\mathcal{I} = \{\boldsymbol{w} \in \mathbb{R}^c : w_1 \leq \ldots \leq w_c\}$ with weights $\boldsymbol{t}$. The simple and elegant pool adjacent violators algorithm (PAVA) can be used (see Robertson et al., 1988).

For Patefield's data, the empirical distribution of the treatment can be shown to be stochastically larger than the empirical distribution of the control; that is, $\hat{F}_2(j) \leq \hat{F}_1(j)$, $j = 1, 2, 3$. The scores that minimize and maximize T^w are $\boldsymbol{w}^{\min} = (0, 1, 1)$ and $\boldsymbol{w}^{\max} = (0, 0.228, 1)$, respectively, with corresponding p-values of 0.1534 and 0.0052. We find that there are, in this straddling case, some scores that produce significance and some others that do not. A graphical representation in terms of $T_{\boldsymbol{w}} = \hat{\rho}$ is given in Figure 2.3.

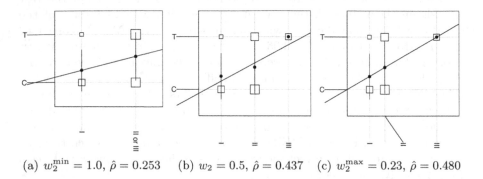

(a) $w_2^{\min} = 1.0$, $\hat{\rho} = 0.253$ (b) $w_2 = 0.5$, $\hat{\rho} = 0.437$ (c) $w_2^{\max} = 0.23$, $\hat{\rho} = 0.480$

Fig. 2.3. Correlation coefficients as a function of $\boldsymbol{w}$.

However, we cannot consider $T_{\boldsymbol{w}}$ with fixed scores $\boldsymbol{w}^{\max} = (0, 0.23, 1)$ because data-snooping bias arises. Gross (1981) suggested that an "analysis based on [...] data-dependent scores may yield procedures that compare favorably to fixed-scores procedures". An adaptive test (Hogg, 1974; Berger and Ivanova, 2002a) based on the test statistic

$$T_{\max} = T_{\boldsymbol{w}^{\max}} = \max(\hat{\rho} : \boldsymbol{w} \in \mathfrak{J}, w_1 < w_c) \qquad (2.4)$$

can be constructed by computing the data-dependent scores $\boldsymbol{w}^{\max}$ at each permutation of the data. For instance, $\boldsymbol{w}^{\max} = (0, 0.41, 1)$ from the contingency table $\{(6, 10, 1); (2, 9, 4)\}$ obtained by exchanging one control value from "same" to "better" and one treatment value from "better" to "same" (that is, an arbitrary bth permutation of the data), obtaining $T_{\max}^*(b) = 0.288$.

For Patefield's data, the permutation distributions of $T_{\boldsymbol{w}} = \hat{\rho}$ by using different scoring systems are displayed in Figure 2.4, and results are given in Table 2.3. We can see that the permutation distributions of $T_{(0, .5, 1)}$ and $T_{\max}$ are rather discrete (there were only 12 different realized values out of 54), making the tests automatically more conservative. Possible solutions are to make use of the mid-p-value (Lancaster, 1961) or a backup statistic (see Cohen et al., 2003).

The critical function of the randomized adaptive test is

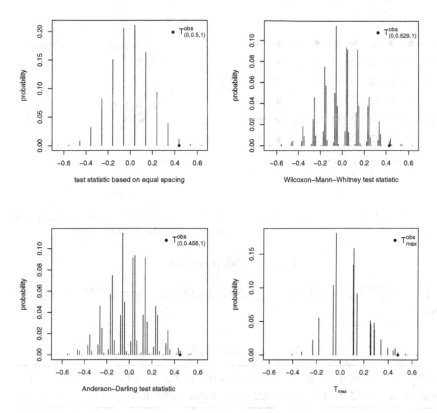

Fig. 2.4. Permutation distributions of $T_w = \hat{\rho}$.

Table 2.3. Results for Patefield's data.

Value System	w	T_w^{obs}	p^{obs}
equal spacing	(0,0.500,1)	0.438	0.0133
midranks	(0,0.529,1)	0.428	0.0133
Anderson-Darling	(0,0.456,1)	0.450	0.0059
adaptive	w^{max}	0.480	0.0073

$$\phi_{T_{max}} = \begin{cases} 1 & \text{if } T_{max} > .342 \\ 0.28 & \text{if } T_{max} = .342 \\ 0 & \text{if } T_{max} < .342 \end{cases}.$$

The rejection region and the conditional power function of the adaptive test are given in Figures 2.5 (a) and (b), respectively. The adaptive test can be viewed as conditionally Bayes with respect to a prior putting its mass on the

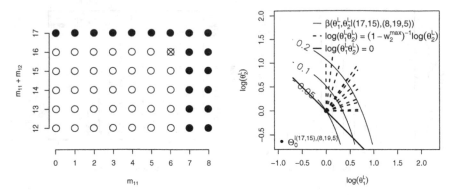

(a) 0.043 level rejection region (nonran-   (b) Contour plot of the conditional power
domized test)   function for the adaptive test

Fig. 2.5. Adaptive test.

set $\{\boldsymbol{\nu} \in \Theta_1^{|(\boldsymbol{n},\boldsymbol{t})} : \log(\theta_1^L \theta_2^L) = \frac{1}{(1-w_2^{\max})} \log(\theta_2^L)\}$, where $w_2^{\max}$ are given in Table 2.4.

Table 2.4. Values of $w_2^{\max}$.

$w_2^{\max}$	0	0.108	0.228	0.398	0.407	0.438	0.526	0.594	0.723	0.789	0.955	1
prob.	0.343	0.037	0.003	0.001	0.022	0.091	0.001	0.006	0.001	0.045	0.0106	0.435

We can see from the difference between conditional powers in Figure 2.6 that the adaptive test distributes the power more uniformly over the entire alternative space.

The power of $T_{\boldsymbol{w}}$ given $\boldsymbol{n} = (17, 15)$ (i.e., the *unconditional power* for fixed sample sizes) is given by

$$\beta(\boldsymbol{\theta}^C | \boldsymbol{n} = (17, 15)) = \mathrm{E}_{\boldsymbol{\theta}^C} [\phi_{T_{\boldsymbol{w}}} | \boldsymbol{n} = (17, 15)], \quad \forall \boldsymbol{\theta}^C = (\theta_1^C, \theta_2^C)^t \in \Theta_0 \cup \Theta_1.$$

We replicate the unconditional power study in Cohen and Sackrowitz (2000) by using the algorithm of Patefield (1981) for generating all the possible tables. We compare the empirical power of Cramér-von Mises (equal spaced scores), WMW, Anderson-Darling, and adaptive tests with the most powerful test (Berger, 1998), which is based on $\lambda = (\nu_1 - \nu_2)/\nu_1 \in \mathcal{R}$, and it can be expressed as

$$T_{\boldsymbol{w}} : \begin{cases} \boldsymbol{w} = (-\lambda/(1-\lambda), 0, 1) & \text{if } \lambda < 0 \\ \boldsymbol{w} = (0, \lambda, 1) & \text{if } \lambda \in [0, 1]. \\ \boldsymbol{w} = (0, 1, 1/\lambda) & \text{if } \lambda > 1 \end{cases}$$

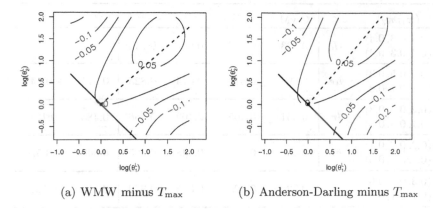

(a) WMW minus $T_{\max}$ (b) Anderson-Darling minus $T_{\max}$

Fig. 2.6. Difference between conditional powers.

Simulations are based on 2000 Monte Carlo replicates with a nominal level of $\alpha = 0.05$, where the row probabilities and the stochastic order relationship (i.e., likelihood ratio "lr", hazard ratio "hr", and simple stochastic order "st") are given in Table 2.5.

2.2.3 Applications with R functions

Here we provide assistance in doing the statistical tests illustrated in Subsection 2.2.2 using the R language. Create Patefield's data with equal-spaced scores $(w_1, w_2, w_3) = (1, 2, 3)$, where X represents class labels and Y the vector of data,

```
> X <- c(rep(1,17),rep(2,15))
> Y <- c(rep(1,6),rep(2,11),rep(1,2),rep(2,8),rep(3,5))
```

and obtain w_2 for the mid-rank and Anderson-Darling scores

```
> mr <- rank(Y)
> w2.mr <- (sort(unique(mr))[2]-min(mr))/(max(mr)-min(mr))
> w2.mr
[1] 0.5294118
> F <- ecdf(Y)
> varF <-sort(unique(F(Y)))*(1-sort(unique(F(Y))))
> w2.ad<-(1/sqrt(varF[1]))/(1/sqrt(varF[1])+1/sqrt(varF[2]))
> w2.ad
[1] 0.4560859
```

Perform the ptest2s function for comparing two independent samples based on Student's t statistic T_w with equal-spaced scores

Table 2.5. Unconditional power comparisons.

$(\pi_1; \pi_2)$	$\overset{\circ}{\lessgtr}$	C-vM	WMW	A-D	$T^{\max}$	MP	λ
$(0.3, 0.3, 0.4)$ $(0.3, 0.3, 0.4)$	H_0	0.033	0.042	0.040	0.040	–	–
$(0.2, 0.5, 0.3)$ $(0.1, 0.1, 0.8)$	hr	0.692	0.814	0.674	0.823	0.915	-0.547
$(0.2, 0.5, 0.3)$ $(0.1, 0.3, 0.6)$	lr	0.391	0.465	0.409	0.456	0.485	0.131
$(0.2, 0.7, 0.1)$ $(0.2, 0.3, 0.5)$	hr	0.430	0.540	0.448	0.695	0.840	-0.526
$(0.3, 0.1, 0.6)$ $(0.1, 0.1, 0.8)$	lr	0.320	0.323	0.345	0.340	0.350	0.792
$(0.3, 0.2, 0.5)$ $(0.1, 0.3, 0.6)$	st	0.198	0.202	0.262	0.279	0.377	1.174
$(0.4, 0.4, 0.2)$ $(0.1, 0.7, 0.2)$	st	0.248	0.358	0.274	0.447	0.627	1.403
$(0.4, 0.5, 0.1)$ $(0.3, 0.1, 0.6)$	hr	0.584	0.634	0.664	0.853	0.960	-0.635
$(0.6, 0.2, 0.2)$ $(0.4, 0.2, 0.4)$	lr	0.265	0.292	0.294	0.295	0.297	0.369
$(0.6, 0.3, 0.1)$ $(0.1, 0.4, 0.5)$	lr	0.940	0.956	0.949	0.948	0.962	0.611
$(0.6, 0.3, 0.1)$ $(0.2, 0.1, 0.7)$	lr	0.947	0.953	0.961	0.974	0.975	0
$(0.4, 0.5, 0.1)$ $(0.4, 0.4, 0.2)$	hr	0.067	0.080	0.103	0.117	0.175	-0.322

```
> source("ptest2s.R")
> set.seed(0)
> B <- 5000
> T <- ptest2s(Y,X,B,"Student")
> p.obs <- sum(T[-1]>=T[1])/(B-1)
> p.obs
[1] 0.01140228
```

and with midrank or Anderson-Darling scores

```
> Y.mr <- Y
> Y.mr[Y.mr==1] <- 0
> Y.mr[Y.mr==2] <- w2.mr
> Y.mr[Y.mr==3] <- 1
```

```
> set.seed(0)
> T <- ptest2s(Y.mr,X,B,"Student")
> p.obs <- sum(T[-1]>=T[1])/(B-1)
> p.obs
[1] 0.01140228
> Y.ad <- Y.mr
> Y.ad[Y.ad==w2.mr] <- w2.ad
> set.seed(0)
> T <- ptest2s(Y.ad,X,B,"Student")
> p.obs <- sum(T[-1]>=T[1])/(B-1)
> p.obs
[1] 0.00620124
```

The adaptive test $T_{\boldsymbol{w}^{\max}}$ can be performed by computing the data-dependent scores $\boldsymbol{w}^{\max}$ at each permutation of the data using the function Tmax:

```
> source("Tmax.R")
> set.seed(0)
> T <- ptest2s(Y,X,B,"Tmax")
> p.obs <- sum(T[-1]>=T[1])/(B-1)
> p.obs
[1] 0.00680136
```

2.2.4 Concordance Monotonicity

Likelihood inference is perhaps the default approach for many statistical models. Recently, there have been debates about the suitability of different test procedures: Perlman and Chaudhuri (2004b) argue in favor of likelihood ratio tests, whereas Cohen and Sackrowitz (2004) argue in favor of the so-called class of directed tests. The likelihood ratio test (LRT) statistic is given by

$$
T_{LR} = 2 \sum_{i=1}^{2} \sum_{j=1}^{c} m_{i,j} \left\{ \log[\hat{\pi}_i^{|H_1}(j)] - \log[\hat{\pi}_i^{|H_0}(j)] \right\},
$$

where $\hat{\pi}_i^{|H_1}(j)$ and $\hat{\pi}_i^{|H_0}(j)$ are the maximum likelihoood (ML) estimates of $\pi_i(j)$ under H_1 and H_0, respectively. When $m_{i,j} > 0$, $i = 1, 2$, $j = 1, \ldots, c$, Dykstra et al. (1996) showed that ML estimates can be expressed in terms of a weighted least squares projection,

$$
\hat{\pi}_1^{|H_1}(j) = \frac{\boldsymbol{m}_1}{n_1} \left\{ \frac{n_1}{n} + \frac{n_2}{n} P_{\frac{\boldsymbol{m}_1}{n_1}} \left(\frac{\boldsymbol{m}_1}{\boldsymbol{m}_2} | \mathfrak{I} \right) \right\},
$$

$$
\hat{\pi}_2^{|H_1}(j) = \frac{\boldsymbol{m}_2}{n_2} \left\{ \frac{n_2}{n} + \frac{n_1}{n} P_{\frac{\boldsymbol{m}_2}{n_2}} \left(\frac{\boldsymbol{m}_2}{\boldsymbol{m}_1} | \mathcal{D} \right) \right\},
$$

where $\mathfrak{I} = \{ \boldsymbol{w} \in \mathcal{R}^c : w_1 \leq \ldots \leq w_c \}$ and $\mathcal{D} = \{ \boldsymbol{w} \in \mathcal{R}^c : w_1 \geq \ldots \geq w_c \}$.

Then, a least favorable null value for the asymptotic distribution of the LRT assigns probability $\frac{1}{2}$ for the first and the last ordinal categories (see Silvapulle and Sen, 2005, Proposition 6.5.1), and

$$\sup_{H_0} \lim_{n \to \infty} \Pr(T_{LR} \geq t | H_0) = \frac{1}{2}(\Pr(\chi^2_{c-2} \geq t) + \Pr(\chi^2_{c-1} \geq t)).$$

To bypass possibly poor asymptotic approximations, mostly for small or unbalanced sample sizes, Agresti and Coull (2002) suggest performing the permutation test based on the LRT statistic. The cell entries in Tables I and II of Table 2.6 represent two permutations of Patefield's data.

Table 2.6. Two permutations of Patefield's data.

Table I	W	S	B	Total
C	5	11	1	17
T	3	8	4	15
	8	19	5	32

Table II	W	S	B	Total
C	0	16	1	17
T	8	3	4	15
	8	19	5	32

Note that Table II is created from Table I by exchanging five control values from "worse" to "same" and five treatment values from "same" to "worse". The LRT statistic for Tables I and II is 2.777 and 22.652, respectively. This seems to contradict intuition because the control performance is improved while simultaneously the treatment is made to perform worse. Then we would expect the p-value to increase. The LRT does not have this property (i.e., is not *concordant monotone*, Cohen and Sackrowitz, 1998), meaning that the p-value decreases if any entry in the first row, say $m_{1,j}$, increases while $m_{1,l}$ decreases for $j < l$, holding all row and column totals fixed.

As an alternative to LRT, Cohen et al. (2003) developed the directed chi-square, which is concordant monotone and is defined as

$$T_{\vec{\chi}^2} = \inf_{u \in \mathcal{A}} \sum_{i=1}^{2} \sum_{j=1}^{c} \frac{\left(u_{i,j} - \frac{n_i t_j}{n}\right)^2}{\frac{n_i t_j}{n}},$$

where $\mathcal{A} = \{u_{1,1} + \ldots + u_{1,j} \geq m_{1,1} + \ldots + m_{1,j}, \sum_{j=1}^{c} u_{i,j} = n_i, u_{1,j} + u_{2,j} = t_j, i = 1, 2, j = 1, \ldots, c\}$. Therefore, the directed chi-squared test rejects H_0 if the minimum of the Pearson chi-square for tables in $\mathcal{A}$ is large. Cohen et al. (2003) showed that a permutationally equivalent formulation is given by

$$T_{\vec{\chi}^2} = \sum_{j=1}^{c} (w_j)^2 t_j,$$

where $\boldsymbol{w} = P_t\left(\frac{\boldsymbol{m_1}}{t} | \mathcal{D}\right)$ and $\mathcal{D} = \{\boldsymbol{w} \in \mathbb{R}^c : w_1 \geq \ldots \geq w_c\}$.

2.2.5 Applications with R functions

Here we provide assistance in doing the statistical tests illustrated in Subsection 2.2.4. Create Table I and Table II representing two permutations of Patefield's data by using equal-spaced scores $(w_1, w_2, w_3) = (1, 2, 3)$, where X represents class labels and Y the vector of data

```
> X <- c(rep(1,17),rep(2,15))
> YI <- c(rep(1,5),rep(2,11),3,rep(1,3),rep(2,8),rep(3,4))
> YII <- c(rep(2,16),rep(3,1),rep(1,8),rep(2,3),rep(3,4))
```

and perform the likelihood ratio test T_{LR} by LRT

```
> source("LRT.R")
>  LRT(YI[X==1],YI[X==2])
[1] 2.783381
>  LRT(YII[X==1],YII[X==2])
[1] 19.4776
```

and the directed chi-squared test $T_{\vec{\chi}^2}$ by DChisq

```
> source("DChisq.R")
>  DChisq(YI,X)
[1] 0.07446918
>  DChisq(YII,X)
[1] 0.07037817
```

Tests based on linear test statistics are also concordant monotone. For example,

```
> source("studT.R")
> studT(YI[X==1],YI[X==2])
[1] 1.348210
> studT(YI[X==1],YI[X==2])
[1] -1.460447
> source("Tmax.R")
> Tmax(YI[X==1],YI[X==2])
[1] 0.2882637
> Tmax(YII[X==1],YII[X==2])
[1] 0.2856531
```

2.2.6 Multiple Testing

In this multiparameter problem, following Roy's (1953) union-intersection principle , it might be possible to look upon the null hypothesis as the intersection of several component hypotheses and the alternative hypothesis as the union of the same number of component alternatives, in symbols

$$H_0 : \boldsymbol{\theta}_1^C = \mathbf{1}_{c-1} \Leftrightarrow \bigcap_{j=1}^{c-1} \{H_{0,j}\} : \bigcap_{j=1}^{c-1} \{\theta_{1j}^C = 1\} ,$$

stating that H_0 is true if all $H_{0,j}$ are true, and

$$H_1 : \boldsymbol{\theta}_1^C \gneqq \mathbf{1}_{c-1} \Leftrightarrow \bigcup_{j=1}^{c-1} \{H_{1,j}\} : \bigcup_{j=1}^{c-1} \{\theta_{1j}^C > 1\} ,$$

stating that H_1 is true if at least one $H_{1,j}$ is true.

To provide an interpretation of this, let us consider the $c-1$ possible 2×2 subtables that can be formed by dichotomizing the column variable: the first column vs. all the rest, the first two columns pooled vs. the others, and so on. Thus $H_{0,j}$ and $H_{1,j}$ define the hypotheses of interest for the jth subtable, $j = 1, \ldots, c-1$ (Table 2.7).

Table 2.7. jth subtable.

	$\leq j$	$> j$	
1	$\sum_{l=1}^{j} m_{1,l}$	$\sum_{l=j+1}^{c} m_{1,l}$	n_1
2	$\sum_{l=1}^{j} m_{2,l}$	$\sum_{l=j+1}^{c} m_{2,l}$	n_2
	$\sum_{l=1}^{j} t_l$	$\sum_{l=j+1}^{c} t_l$	n

For testing $H_{0,j}$ against $H_{1,j}$, we may consider Fisher's test statistic $T_j = \sum_{l=1}^{j} m_{1,l}$ or its standardized formulation

$$T_j = \left(\frac{n_1 n_2}{n^2/(n-1)} \right) \frac{\hat{F}_1(j) - \hat{F}_2(j)}{(\hat{F}(j)[1 - \hat{F}(j)])^{\frac{1}{2}}}.$$

For any $K \subseteq \{1, \ldots c-1\}$, let $H_{0,K} : \bigcap_{j \in K} \{H_{0,j}\}$ denote the hypothesis that all $H_{0,j}$ with $j \in K$ are true. The closure method of Marcus et al. (1976) allows strong control of FWE if we know how to test each intersection hypothesis $H_{0,K}$. Let T_K be a test statistic for $H_{0,K}$ that can be a function of test statistics T_j or p-values p_j. For instance, by using the standardized Fisher statistic, the "sum-T" combined test $T_K = \sum_{j \in K} T_j$ yields the Anderson-Darling statistic, whereas by using $T_j = \hat{F}_1(j) - \hat{F}_2(j)$, the "max-T" combined test $T_K = \max_{j \in K} T_j$ yields the Smirnov statistic.

For the analysis of Patefield's data, by applying Fisher's exact tests, we obtain $p_1^{obs} = 0.1536$ and $p_2^{obs} = 0.0149$. Depending on the combined test used to test H_0, obtaining p-value p^{obs}, from the closed testing principle we have $p_1^{adj} = \max(0.1536, p^{obs})$ and $p_2^{adj} = \max(0.0149, p^{obs})$. When $p^{obs} \leq \alpha$, as happens with the tests considered in Subsection 2.2.2, the "individual" hypothesis $H_{0,2} : \theta_2^C = 1$ can be rejected while controlling the FWE, supporting the alternative $H_{1,2} : \theta_2^C > 1$.

2.3 Independent Binomial Samples: $r \times 2$ Contingency Tables

In a typical dose-response study, several increasing doses of a treatment are randomly assigned to the subjects, with each subject receiving only one dose throughout the study. We discuss the case of a binary response variable Y with a single regressor X having r ordered levels. Let $Y_1, \ldots, Y_r$ be r independent binomial variables with $Y_i \sim \text{Binomial}(n_i, \pi_i)$, where the probability of "success" is $\pi_i := \pi_i(2) = \Pr(Y = 2 | X = i)$. We are often interested in detecting inequalities between the parameters π_i, $i = 1, \ldots, r$.

Graubard and Korn (1987) consider a prospective study of maternal drinking and congenital malformations. After the first three months of pregnancy, the women in the sample completed a questionnaire about alcohol consumption (average number of drinks per day). Following childbirth, observations were recorded on the presence or absence of congenital sex organ malformations. The data are displayed in Table 2.8.

Table 2.8. Maternal drinking and congenital malformations data.

X	Alcohol Consumption	Malformation Absent	Present	
1	0	17,066	48	17,114
2	< 1	14,464	38	14,502
3	1–2	788	5	793
4	3–5	126	1	127
5	≥ 6	37	1	38
		32,481	93	32,574

The goal is to test for a dose-response relationship. For example, when investigating a dose-response relationship of the form $\text{logit}(\pi_i) = \gamma + \beta d_i$, one would typically have in mind a biologically or clinically meaningful slope, say β, above which one could claim the existence of a trend in the data. Specifically, it is of interest to test against a *simple order restriction*,

$$H_0 : \pi_1 = \ldots = \pi_r \quad \text{against} \quad H_1 : \pi_1 \leq \ldots \leq \pi_r, \qquad (2.5)$$

with at least one strict inequality. An efficient test of the null hypothesis is the Cochran-Armitage test of trend (Cochran, 1954; Armitage, 1955), in which the test statistic is

$$T_d = \sum_{i=1}^{r-1} m_{i,2} d_i,$$

where the d_i's are prespecified scores that may correspond to doses in a dose-response setting. It is known (Agresti, 2002, pp. 181−182) that the Cochran-Armitage statistic is equivalent to the score statistic for testing $H_0 : \beta = 0$ in the linear logit model. Cochran (1954) noted that "any set of scores gives a valid test, provided that they are constructed without consulting the results of the experiment. If the set of scores is poor, in that it badly distorts a numerical scale that really does underlie the ordered classification, the test will not be sensitive. The scores should therefore embody the best insight available about the way in which the classification was constructed and used." Ideally, the scale is chosen by a consensus of experts, and subsequent interpretations use that same scale. When uncertain about this choice, the adaptive scores used in Subsection 2.2.2 may be considered.

Alcohol consumption, measured as the average number of drinks per day, is an ordinal explanatory variable. This groups a naturally continuous variable, and we first use the scores $d = (0, .5, 1.5, 4, 7)$, the last score being somewhat arbitrary. For this choice, the p-value is 0.014. By contrast, for the equally spaced row scores, $d = (1, 2, 3, 4, 5)$, giving a much weaker conclusion ($p = 0.104$). Midrank scores yield an even weaker conclusion ($p = 0.319$). Why does this happen? Adjacent categories having relatively few observations necessarily have similar midranks. This scoring scheme treats the alcohol consumption level 1−2 drinks as much closer to consumption level ≥ 6 drinks than to consumption level 0 drinks. This seems inappropriate since it is usually better to select scores that reflect distances between doses. However, by using the adaptive scores $d = w^{\max}$, the p-value is 0.022, supporting the adaptive test when the choice of scores is uncertain.

Peddada et al. (2001) consider a study investigating the effects of several treatments on the reproductive condition of the redbacked salamander (*Plethodon cinereus*). Female salamanders were randomly assigned to either the control or one of three treatment groups. The treatments consisted of injections of either follicle-stimulating hormone ($i = 2$), luteinizing hormone ($i = 3$), or, for animals in the control group ($i = 1$), saline solution. The remaining treatment group ($i = 4$) was fed exactly twice the amount of food as salamanders in all other groups. The reproductive condition of each animal was later evaluated by measuring the size of the ova through the abdominal wall of the animal. If the ova were larger than 2 mm, then the animal was declared to be in a reproductive condition. Data are displayed in Table 2.9.

The hypothesis of interest is whether the salamanders in the treatment groups had a greater probability of being in reproductive condition than those in the control group. No ordering was hypothesized between treatment groups, and hence we wish to test against a *simple tree order restriction*

Table 2.9. Reproductive condition of the redbacked salamander.

	Nonreproductive	Reproductive	Total
1	4	9	13
2	8	4	12
3	7	6	13
4	1	13	14
	20	22	42

$$H_0 : \pi_1 = \ldots = \pi_r \quad \text{against} \quad H_1 : \bigcup_{i=1}^{r-1} \{\pi_1 < \pi_{i+1}\}. \tag{2.6}$$

In the case of rejection of the global null hypothesis that none of the treatments is an improvement over the control, answering the question "Is there any evidence of the treatment effect?" one usually wants to know which of the treatments show a significant difference, answering the more specific question "For which treatments is the response larger than the response in the control group?"; that is, to test simultaneously the hypotheses

$$H_{0,i} : \pi_1 = \pi_{i+1} \quad \text{against} \quad H_{1,i} : \pi_1 < \pi_{i+1}, \qquad i = 1, \ldots, r-1. \tag{2.7}$$

A multiple comparison procedure can be used for this purpose. No type I error should be made in any of these comparisons because otherwise a treatment that is actually inferior to the control may be recommended. Thus, in this case, strong control of the FWE is required.

For any $K \subseteq \{1, \ldots r-1\}$, let $H_{0,K} : \bigcap_{i \in K} H_{0,i}$ denote the hypothesis that all $H_{0,i}$ with $i \in K$ are true. Note that for testing the global null hypothesis H_{0,K_0} with $K_0 = \{1, \ldots, r-1\}$, all permutations of the observations among the r groups are equally likely. However, for testing the intersection hypothesis $H_{0,K}$, we consider only the permutations that, under that hypothesis, become equally likely. In particular, for each hypothesis $H_{0,i}$, one has to permute only within the control and the $(i+1)$th treatment. Thus we should consider a closed testing procedure that uses the valid permutations depending on the intersection hypothesis under testing. For testing $H_{0,i}$, we consider as test statistics

$$T_i = \frac{\hat{\pi}_{i+1} - \hat{\pi}_1}{\left[\left(\frac{n_1 \hat{\pi}_1 + n_{i+1} \hat{\pi}_{i+1}}{n_1 + n_{i+1}} \right) \left(1 - \frac{n_1 \hat{\pi}_1 + n_{i+1} \hat{\pi}_{i+1}}{n_1 + n_{i+1}} \right) \right]^{\frac{1}{2}}}, \qquad i = 1, \ldots, r-1,$$

where $\hat{\pi}_i = m_{i2}/n_i$, and for testing H_K the "max-T" combined test statistic $T_K = \max(T_i : i \in K)$. Results are given in the following diagram to better illustrate the closed testing method. The result indicates that no individual hypothesis can be rejected at a nominal level $\alpha = 5\%$.

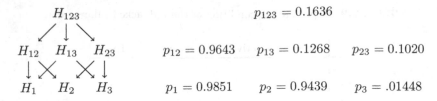

$$H_{123} \qquad\qquad\qquad p_{123} = 0.1636$$

$$H_{12} \quad H_{13} \quad H_{23} \qquad p_{12} = 0.9643 \quad p_{13} = 0.1268 \quad p_{23} = 0.1020$$

$$H_1 \quad H_2 \quad H_3 \qquad p_1 = 0.9851 \quad p_2 = 0.9439 \quad p_3 = .01448$$

Note that in all examples there is prior belief in the shape of the exposure-outcome curve. The prior belief relates to a restricted alternative to the "no effect" hypothesis. For instance, in the salamander example, *a priori expectations* were that all three treatment groups would result in increased ova development compared with a control.

2.3.1 Applications with R functions

This paragraph provides assistance in using the statistical methods illustrated in Section 2.3. Create the malformations data with equal-spaced scores, where X represents class labels and Y the vector of data,

```
> X <- c(rep(1,17114),rep(2,14502),rep(3,793),rep(4,127),
+ rep(5,38))
> Y<-c(rep(0,17066),rep(1,48),rep(0,14464),rep(1,38),
+ rep(0,788),rep(1,5),rep(0,126),rep(1,1),rep(0,37),rep(1,1))
```

and perform the ptest2s function based on T_d by switching the input arguments X and Y; with equally spaced, scores, we obtain

```
> source("ptest2s.R")
> set.seed(0)
> B <- 1000
> T <- ptest2s(X,Y,B,"Student")
> p.obs <- sum(T[-1]>=T[1])/(B-1)
> p.obs
[1] 0.1041041
```

whereas with $d = (0, .5, 1.5, 4, 7)$ and midrank scores, we obtain

```
> set.seed(0)
> X.d <- c(rep(0,17114),rep(.5,14502),rep(1.5,793),
+ rep(4,127),rep(7,38))
> T <- ptest2s(X.d,Y,B,"Student")
> p.obs <- sum(T[-1]>=T[1])/(B-1)
> p.obs
[1] 0.01401401
> set.seed(0)
> X.mr <- rank(X)
> T <- ptest2s(X.mr,Y,B,"Student")
```

```
> p.obs <- sum(T[-1]>=T[1])/(B-1)
> p.obs
[1] 0.3193193
```

Finally, by using adaptive scores $d = w^{\max}$, we obtain

```
> source("Tmax.R")
> set.seed(0)
> T <- ptest2s(X,Y,B,"Tmax")
> p.obs <- sum(T[-1]>=T[1])/(B-1)
> p.obs
[1] 0.02202202
```

To set up the redbacked salamander data, type

```
> X <- c(rep(1,13),rep(2,12),rep(3,13),rep(4,14))
> Y <- c(rep(0,4),rep(1,9),rep(0,8),rep(1,4),
+ rep(0,7),rep(1,6),rep(0,1),rep(1,13))
```

To test the global null hypothesis $H_{0,\{1,2,3\}} : \bigcap_{i=1}^{3}\{\pi_1 = \pi_{i+1}\}$, perform the ptestRs function based on the combined statistic $T_{\{1,2,3\}} = \max(T_1, T_2, T_3)$

```
> source("ptestRs.R")
> set.seed(0)
> B <- 5000
> T123 <- ptestRs(Y,X,B,combi="max")
> p.obs <- sum(T123[-1]>=T123[1])/(B-1)
> p.obs
[1] 0.1672334
```

and for the intersection hypotheses $H_{0,\{1,2\}}$, $H_{0,\{1,3\}}$, and $H_{0,\{2,3\}}$

```
> T12 <- ptestRs(Y[X!=4],X[X!=4],B,combi="max")
> sum(T12[-1]>=T12[1])/(B-1)
[1] 0.9643929
> T13 <- ptestRs(Y[X!=3],X[X!=3],B,combi="max")
> sum(T13[-1]>=T13[1])/(B-1)
[1] 0.1254251
> T23 <- ptestRs(Y[X!=2],X[X!=2],B,combi="max")
> sum(T23[-1]>=T23[1])/(B-1)
[1] 0.1028206
```

Finally, to test $H_{0,1}$, $H_{0,2}$ and $H_{0,3}$, perform the two-sample comparisons by using the ptest2s function

```
> T1 <- ptest2s(Y[X!=3&X!=4],X[X!=3&X!=4],B,"Student")
> sum(T1[-1]>=T1[1])/(B-1)
[1] 0.9867974
> T2 <- ptest2s(Y[X!=2&X!=4],X[X!=2&X!=4],B,"Student")
> sum(T2[-1]>=T2[1])/(B-1)
```

```
[1] 0.9461892
> T3 <- ptest2s(Y[X!=2&X!=3],X[X!=2&X!=3],B,"Student")
> sum(T3[-1]>=T3[1])/(B-1)
[1] 0.1386277
```

2.4 Comparison of Several Treatments when the Response is Ordinal: $r \times c$ Contingency Tables

Table 2.10 displays data appearing in Chuang-Stein and Agresti (1997). Five ordered categories ranging from "death" to "good recovery" describe the clinical outcome of patients who experienced trauma. In the literature on critical care, these five categories are often called the Glasgow Outcome Scale (GOS). We have four treatment groups: three intravenous doses for the medication (low, medium, and high) and a vehicle infusion serving as the control.

Table 2.10. Glasgow Outcome Scale.

Treatment group	X	Death	Vegetative state	Major disability	Minor disability	Good recovery	Total
Placebo	1	59	25	46	48	32	210
Low dose	2	48	21	44	47	30	190
Medium dose	3	44	14	54	64	31	207
High dose	4	43	4	49	58	41	195
		194	64	193	217	134	802

Investigation of a dose-response relationship is of primary interest in many drug-development studies. Here the outcome of interest is measured at several (increasing) dose levels, among which there is a control group. One study objective was to determine whether a more favorable GOS outcome tends to occur as the dose increases; that is, testing

$$H_0 : Y_1 \overset{d}{=} \ldots \overset{d}{=} Y_r \quad \text{against} \quad H_1 : Y_1 \overset{st}{\leq} \ldots \overset{st}{\leq} Y_r$$

with at least one "$\overset{st}{\leq}$". Note that the dose-response curve is assumed to be monotone; i.e., the GOS increases as the dose level increases.

Other questions usually asked in dose-response studies are: "For which doses is the response higher from the response in the control group?"; that is, testing the many-to-one comparisons

$$H_{0,i} : Y_1 \overset{d}{=} Y_{i+1} \quad \text{against} \quad H_{1,i} : Y_1 \overset{st}{\leq} Y_{i+1}, \qquad i = 2, \ldots, r,$$

or "What are the strict inequalities in the stochastic ordering relationship?", that is, testing all pairwise comparisons

$$H_{0,(i,i')} : Y_i \stackrel{d}{=} Y_{i'} \quad \text{against} \quad H_{1,(i,i')} : Y_i \stackrel{st}{\lesssim} Y_{i'}, \qquad i < i'.$$

Gatekeeping procedures (see Dmitrienko and Tamhane, 2007, for an overview) have become popular in recent years as they provide a convenient way to handle logical relationships between multiple objectives that clinical trials are often required to address. In a gatekeeping strategy, the k null hypotheses are divided into h ordered families $\mathcal{F}_l$, $l = 1, \ldots, h$. Generally, familywise error rate control at a designated level α is desired for the family of all k hypotheses.

Westfall and Krishen (2001) proposed procedures for the *serial* gatekeeping problem in which the hypotheses in $\mathcal{F}_{l+1}$ are tested if and only if all hypotheses in $\mathcal{F}_l$ are rejected. Dmitrienko et al. (2003) proposed procedures for the *parallel* gatekeeping problem in which the hypotheses in $\mathcal{F}_{l+1}$ are tested if at least one hypothesis in $\mathcal{F}_l$ is rejected.

For the many-to-one comparisons, the serial gatekeeping procedure may exploit the hierarchy of the stochastic ordering relationship $Y_1 \stackrel{st}{\leq} \ldots \stackrel{st}{\leq} Y_r$ by starting from the comparison between the highest dose and the control to the comparison between the lowest dose and the control. Here, the ordered families are simply $\mathcal{F}_1 = \{H_{0,r}\}, \ldots, \mathcal{F}_{r-1} = \{H_{0,2}\}$. The procedure stops at the dose level where the null hypothesis is not rejected at the nominal level $\alpha = 5\%$.

For the pairwise comparisons, a parallel gatekeeping procedure may exploit the distance $(i, i') = i' - i$ in the stochastic ordering relationship by testing first the hypothesis $H_{(1,r)}$ comparing the highest dose with the control and, if rejected, both $H_{(1,r-1)}$ and $H_{(2,r)}$, and if at least one is rejected, the three hypotheses $H_{0,(1,r-2)}$, $H_{0,(2,r-1)}$, and $H_{0,(3,r-2)}$, and so on. Here the ordered families are $\mathcal{F}_1 = \{H_{0,(1,r)}\}, \mathcal{F}_2 = \{H_{0,(1,r-1)}, H_{0,(2,r)}\}, \ldots, \mathcal{F}_l = \{H_{0,(i,i')} : i - i' = r - l\}, \ldots, \mathcal{F}_{r-1} = \{H_{0,(i,i+1)}, i < i'\}$. In this procedure, the first $r - 2$ families are tested using the Bonferroni single-step adjustment that tests $\mathcal{F}_l$ at level $\alpha \gamma_l$. The family $\mathcal{F}_{r-1}$ is tested at level $\alpha \gamma_l$ using Holm's stepdown adjustment. Here γ_l is the so-called rejection gain factor for $\mathcal{F}_l$, given by $\gamma_1 = 1$, $\gamma_l = \prod_{j=1}^{l-1} \left(\frac{\text{rejected}(\mathcal{F}_j)}{\text{cardinality}(\mathcal{F}_j)} \right)$, where "rejected($\mathcal{F}_j$)" is the number of rejected hypotheses in $\mathcal{F}_j$; thus γ_l is the product of the proportions of rejected hypotheses in $\mathcal{F}_1$ through $\mathcal{F}_{l-1}$. If no hypotheses are rejected in some family $\mathcal{F}_l$, then $\gamma_j = 0$ for all $j > l$, and all hypotheses in $\mathcal{F}_j$ for $j > l$ are automatically accepted. On the other hand, if all hypotheses are rejected in $\mathcal{F}_1$ through $\mathcal{F}_{l-1}$, then $\gamma_l = 1$ and thus a full α level is used to test $\mathcal{F}_l$, no part of α being used up by the rejected hypotheses ("use it or lose it" principle).

To illustrate the implementation of gatekeeping procedures, consider the GOS example. Raw p-values for the six hypotheses computed from a two-sample T_w test (with equally spaced or adaptive scores) are displayed in Table 2.11.

Table 2.11. Raw p-values for GOS data.

	$H_{0,(1,4)}$	$H_{0,(1,3)}$	$H_{0,(2,4)}$	$H_{0,(1,2)}$	$H_{0,(2,3)}$	$H_{0,(3,4)}$
Equally spaced	0.0026	0.0282	0.0194	0.2819	0.1206	0.1666
Adaptive	0.0018	0.0220	0.0162	0.4563	0.1258	0.1466

For many-to-one comparisons, by the serial gatekeeping procedure we reject at $\alpha = 5\%$ both the hypotheses $H_{0,(1,4)}$ and $H_{0,(1,3)}$ but not the comparison between the lowest dose and control; that is, $H_{0,(1,2)}$. Note that by performing Holm's procedure, with adaptive scores we reject both $H_{0,(1,4)}$ and $H_{0,(1,3)}$ ($p_{(1,4)} = 0.0018 < \alpha/3$ and $p_{(1,3)} = 0.0220 < \alpha/2$) but with equally spaced scores we can reject $H_{0,(1,4)}$ only ($p_{(1,3)} = 0.0282 > \alpha/2$).

For the six pairwise comparisons, with Holm's procedure we reject only the hypothesis $H_{0,(1,4)}$. However, because there are logical implications among the hypotheses and alternatives, Holm's procedure can be improved to obtain a further increase in power (Shaffer, 1986). By considering all possible configurations of true and false hypotheses, all six hypotheses may be true at the first step, but because the hypothesis $H_{0(1,4)}$ is rejected ($p_{(1,4)} \leq \alpha/6$), at least three must be false since if any two distributions differ, at least one of them must differ from the remaining ones. By exploiting logical implications and using p-values from tests based on adaptive scores, we can also reject $H_{0,(2,4)}$ ($p_{(2,4)} = 0.0162 < \alpha/3$). By performing the parallel gatekeeping procedure, with adaptive scores we reject at $\alpha = 5\%$ all the hypotheses in the families $\mathcal{F}_1 = \{H_{0,(1,4)}\}$ and $\mathcal{F}_2 = \{H_{0,(1,3)}, H_{0,(2,4)}\}$ but none of the hypotheses in the family $\mathcal{F}_3 = \{H_{0,(1,2)}, H_{0,(2,3)}, H_{0,(3,4)}\}$, whereas with equally spaced scores we reject only the hypotheses $H_{0,(1,4)}$ and $H_{0,(2,4)}$ because $p_{0,(2,3)} = 0.0282 > \alpha/2$.

3

Multivariate Ordinal Data

3.1 Introduction

Assessing the risks and benefits of a treatment is more comprehensive and sensitive when several variables are considered simultaneously rather than ignoring some or analyzing them separately. For instance, the large number and variety of possible manifestations of a dose effect in a subject's clinical response usually necessitates that several endpoints be observed jointly to avoid missing any crucial effects or interactions.

Many assessment instruments used in the evaluation of toxicity, safety, pain, or disease progression consider multiple ordinal endpoints to fully capture the presence and severity of treatment effects. Contingency tables underlying these correlated responses are often sparse and imbalanced, rendering asymptotic results unreliable or model fitting prohibitively complex without overly simplifying assumptions. The statistical analysis of these data structures is challenging, first because underlying contingency tables for the multivariate categorical responses are very sparse and imbalanced and second because associations of various degrees among the endpoints may mask or enhance effects if not properly taken into account.

We discuss the case comparing two treatments (doses) based on observing for each subject k ordinal variables with possibly different numbers of categories. Let $\boldsymbol{Y}_i = (Y_{i1}, \dots, Y_{ik})^t$ be the multivariate response at dose $i = 1, 2$, where Y_{ih} is ordinal with $c_h \geq 2$ categories, $h = 1, \dots, k$. Suppose we have a total of $n = n_1 + n_2$ subjects randomly assigned to the two doses, such that $\boldsymbol{Y}_{11}, \dots, \boldsymbol{Y}_{1n_1}$ are n_1 i.i.d. observations from a distribution $\pi_1(j_1, \dots, j_k)$ and, independently, $\boldsymbol{Y}_{21}, \dots, \boldsymbol{Y}_{2n_2}$ are n_2 i.i.d. observations from a distribution $\pi_2(j_1, \dots, j_k)$, where $\pi_i(j_1, \dots, j_k)$ denotes the joint probabilities $\Pr(Y_{i1} = j_1, \dots, Y_{ik} = j_k)$, $j_h \in \{1, \dots, c_h\}$ at dose i. To investigate a possible dose effect, we initially set up the null hypothesis

$$H_0 : \boldsymbol{Y}_1 \overset{d}{=} \boldsymbol{Y}_2, \tag{3.1}$$

D. Basso et al., *Permutation Tests for Stochastic Ordering and ANOVA*, Lecture Notes in Statistics, 194, DOI 10.1007/978-0-387-85956-9_3,
© Springer Science+Business Media, LLC 2009

where "$\overset{d}{=}$" means "equal in distribution" (i.e., $\pi_1(j_1,\ldots,j_k) = \pi_2(j_1,\ldots,j_k)$ for all $(j_1,\ldots,j_k) \in \{1,\ldots,c_1\} \times \ldots \times \{1,\ldots,c_k\}$) against the one-sided alternative that the $\boldsymbol{Y}_2$ distribution is stochastically larger and not equal to the $\boldsymbol{Y}_1$ distribution,

$$H_1 : \boldsymbol{Y}_2 \overset{st}{\gneqq} \boldsymbol{Y}_1. \tag{3.2}$$

The following multivariate generalization of stochastic order was considered first by Lehmann (1955) and is also given in Marshall and Olkin (1979).

Definition 3.1. *The k-dimensional random vector $\boldsymbol{Y}_2$ is said to dominate $\boldsymbol{Y}_1$ according to the multivariate stochastic order, or $\boldsymbol{Y}_2$ is stochastically larger than $\boldsymbol{Y}_1$, written $\boldsymbol{Y}_2 \overset{st}{\geq} \boldsymbol{Y}_1$, if*

$$E[g(\boldsymbol{Y}_2)] \geq E[g(\boldsymbol{Y}_1)] \tag{3.3}$$

for all functions $g : \mathbb{R}^k \to \mathbb{R}$ that are increasing in each argument and have finite expectations.

Note that $\boldsymbol{Y}_2 \overset{st}{\geq} \boldsymbol{Y}_1$ implies both order of the c.d.f.'s $\Pr(Y_{11} \leq j_1,\ldots,Y_{1k} \leq j_k) \geq \Pr(Y_{21} \leq j_1,\ldots,Y_{2k} \leq j_k)$ (i.e., smaller or equal values are more likely to occur under the first dose) and order of the survival functions $\Pr(Y_{11} > j_1,\ldots,Y_{1k} > j_k) \leq \Pr(Y_{21} > j_1,\ldots,Y_{2k} > j_k)$ (i.e., larger values are more likely to occur under the second dose). As is natural, we assume two independent multinomial distributions $(n_i, \{\pi_i(j_1,\ldots,j_k)\})$ for the counts in each of the two tables of size $c_1 \times \cdots \times c_k$ that cross-classify the n_i multivariate responses at dose i. It is clear that if $\boldsymbol{Y}_2 \overset{st}{\geq} \boldsymbol{Y}_1$, then $Y_{h2} \overset{st}{\geq} Y_{1h}$ for each $h = 1,\ldots,k$. But the converse is false because the marginal distributions do not determine the joint distribution uniquely unless, of course, the components are independent.

The multivariate setting poses more difficulties than the univariate one because multivariate estimators cannot generally be recast as solutions to standard isotonic regression problems. In addition, once derived, numerical evaluation of these estimators is extremely difficult. Sampson and Whitaker (1989) derived the maximum likelihood estimates for this problem, and Lucas and Wright (1991) proposed the likelihood ratio test. However, estimating the large number of parameters via ML is impossible for sparse and/or high-dimensional data because of the many empty cells.

Consider the case of $k = 2$, so that $\boldsymbol{Y}_1$ and $\boldsymbol{Y}_2$ are two discrete bivariate random vectors with a common support on a $c_1 \times c_2$ lattice $\mathcal{L}_{c_1 \times c_2}$ and probabilities $\pi_i(j_1,j_2) = \Pr(Y_{1i} = j_1, Y_{2i} = j_2)$, $i = 1, 2$, $j_1 = 1,\ldots,c_1, j_2 = 1,\ldots,c_2$. Also let $n_i(j_1,j_2)$ be the number of subjects receiving treatment i and resulting in outcome (j_1,j_2). The following definitions (Sampson and Singh, 2002) are needed in what follows.

Definition 3.2. *We have a matrix partial order $\preceq_M$ on $\boldsymbol{Y}$ if $(j_1,j_2) \preceq_M (j_1',j_2') \Leftrightarrow j_1 \leq j_1'$ and $j_2 \leq j_2'$.*

Definition 3.3. *A subset U of $\mathcal{L}_{c_1 \times c_2}$ is called an upper set if $(j_1, j_2) \in U$, $(j_1', j_2') \in \mathcal{L}_{c_1 \times c_2}$, and $(j_1, j_2) \preceq_M (j_1', j_2') \Rightarrow (j_1', j_2') \in U$.*

$\mathcal{L}_{c_1 \times c_2}$ is a trivial upper set, as is $\emptyset$. An upper set other than $\mathcal{L}_{c_1 \times c_2}$ and $\emptyset$ is called a nontrivial upper set. An alternative formulation of stochastic ordering (see Marshall and Olkin, 1979, prop. 17.B2) equivalent to (3.3) is

$$\Pr\{\mathbf{Y}_1 \in U\} \leq \Pr\{\mathbf{Y}_2 \in U\}, \qquad \forall U \in \mathcal{U},$$

where $\mathcal{U}$ is the class of all $\binom{c_1+c_2}{c_1}$ upper sets. The testing problem becomes

$$H_0 : \bigcap_{\forall U \in \mathcal{U}} \left\{ \sum_{(j_1,j_2) \in U} \pi_1(j_1, j_2) = \sum_{(j_1,j_2) \in U} \pi_2(j_1, j_2) \right\}$$

against

$$H_1 : \bigcup_{\forall U \in \mathcal{U}} \left\{ \sum_{(j_1,j_2) \in U} \pi_1(j_1, j_2) < \sum_{(j_1,j_2) \in U} \pi_2(j_1, j_2) \right\}.$$

For example, when $c_1 = c_2 = 2$, we have $\binom{4}{2} = 6$ upper sets, namely $\mathcal{U} = \{U_\emptyset = \emptyset, U_1 = \{(2,2)\}, U_2 = \{(2,2), (2,1)\}, U_3 = \{(2,2), (1,2)\}, U_4 = \{(2,2), (2,1), (1,2)\}, U_{\mathcal{L}_{2\times2}} = \{(2,2), (2,1), (1,2), (1,1)\}\}$.

In principle, we are able to test simultaneously $H_U : \sum_{(j_1,j_2) \in U} \pi_1(j_1, j_2) = \sum_{(j_1,j_2) \in U} \pi_2(j_1, j_2)$ against $H_U' : \sum_{(j_1,j_2) \in U} \pi_1(j_1, j_2) < \sum_{(j_1,j_2) \in U} \pi_2(j_1, j_2)$ for all possible nontrivial $U \in \mathcal{U}$ by using Fisher's exact tests based on 2×2 tables and combine them to get an overall result about H_0.

	$(j_1, j_2) \in U$	$(j_1, j_2) \notin U$	
1	$\sum_{(j_1,j_2) \in U} n_1(j_1, j_2)$	$\sum_{(j_1,j_2) \notin U} n_1(j_1, j_2)$	n_1
2	$\sum_{(j_1,j_2) \in U} n_2(j_1, j_2)$	$\sum_{(j_1,j_2) \notin U} n_2(j_1, j_2)$	n_2
	$\sum_{i=1}^{2} \sum_{(j_1,j_2) \in U} n_i(j_1, j_2)$	$\sum_{i=1}^{2} \sum_{(j_1,j_2) \notin U} n_i(j_1, j_2)$	n

Consider the data in Table 3.1 taken from Sampson and Singh (2002). This example arises from a double-blinded randomized clinical trial where both the physician and the patient at the end of the trial are asked to evaluate on a multipoint ordinal scale the patient's response to treatment in comparison with their condition at the start of the trial. Such ratings are usually termed global ratings, where each evaluation is on the scale "worse", "no difference", "better", "much better". The purpose of the trial was to demonstrate superiority of the experimental treatment with respect to the control treatment.

In the analysis of such data, treatments are usually compared by evaluating separately the physician's and the patient's judgments. Due to the complexity

Table 3.1. Physician's and patient's global ratings.

Control						Treatment				
Y_1/Y_2	1	2	3	4		Y_1/Y_2	1	2	3	4
1	7	48	38	1		1	5	9	13	9
2	6	17	33	6		2	4	11	35	14
3	1	10	21	6		3	1	11	45	15
4	0	0	3	3		4	0	3	14	11

of such an analysis, it is less usual to see an analysis that compares treatments based simultaneously on both the physician's and the patient's global ratings. We consider $\binom{8}{4} - 2 = 68$ nontrivial upper sets, and by combining the corresponding Fisher exact tests, we obtain a highly significant p-value equal to 0.0001. However, for higher dimensional data, it becomes increasingly difficult to compute all the possible comparisons. For instance, with $k = 3$ and $c = 4$, we have 232,846 nontrivial upper sets, and thus a different solution is needed. Moreover, in the case of rejection of the null hypothesis that the treatment is an improvement over the control, one usually wants to know which of the component variables (if any) show a statistically significant difference.

One collection of endpoints designed to evaluate neurophysiological effects in animals after exposure to a toxin is a biological screening assay composed of roughly 25 endpoints, termed the Functional Observational Battery (FOB) (Moser, 1989; see also the United States Environmental Protection Agency's guideline 40CFR 798.6050). The FOB tries to group endpoints by a common domain, each domain describing a possibly distinct neurological function. Table 3.2 shows the structure of the FOB and displays data from a study designed to measure the presence and severity of neurotoxic effects in animals after exposure to perchlorethylene (PERC), a chemical used in the dry cleaning industry (and suspected of leading to an unusual concentration of leukemia cases in the city of Woburn, MA; see Lagakos et al., 1986). In the study, a total of 40 animals were randomly assigned to either the placebo or four exposure levels of PERC, with 8 animals per dose group, and the FOB was administered at several time intervals. Each animal was evaluated on 25 endpoints, classified into six domains, and the data were converted to a scale from 1 to 4, where a score of 1 indicates absence of the corresponding adverse effect and a score of 4 denotes the most severe reaction (Moser et al., 1995). Table 3.2 summarizes the result of the FOB at the time of peak effect, 4 hours into exposure, for the placebo and the 1.5 g/kg exposure level.

Many similar comprehensive batteries of tests exist in other fields, such as the presence and severity of several adverse events (which are also usually grouped by body function according to some dictionary) in drug safety studies or various assessment scales for medical diagnoses of pain or diseases such as Alzheimer's or Parkinson's. The statistical analysis of these data structures

Table 3.2. Marginal counts of severity scores for adverse effects at two exposure levels observed from the FOB study. The last three columns give a simplistic analysis treating the scores as normal, referring to the regular t statistic and corresponding raw and Bonferroni-Holm adjusted p-values. The quoted domain test statistic T is based on O'Brien's (1984) method applied to all endpoints within a domain, and p-values adjusted via Bonferroni-Holm methods.

Domain	Endpoint	Exposure								t	raw	adj.
		0 g/kg				1.5 g/kg						
		1	2	3	4	1	2	3	4			
Autonomic	Lacrimation	8	0	0	0	5	0	3	0	2.05	0.030	0.626
$T = 0.93$	Salivation	8	0	0	0	8	0	0	0	0.00	0.500	1.000
$p_{adj} = 0.404$	Pupil	7	0	1	0	5	0	3	0	1.13	0.139	1.000
	Defecation	7	0	0	1	7	1	0	0	−0.63	0.731	1.000
	Urination	4	3	1	0	6	1	1	0	−0.67	0.744	1.000
Sensorimotor	Approach	8	0	0	0	4	0	3	1	2.55	0.011	0.253
$T = 2.08$	Click	7	0	1	0	7	0	1	0	0.00	0.500	1.000
$p_{adj} = 0.185$	Tail pinch	6	0	2	0	5	0	3	0	0.51	0.309	1.000
	Touch	8	0	0	0	6	0	0	2	1.53	0.074	1.000
CNS excitability	Handling	6	2	0	0	4	4	0	0	1.00	0.167	1.000
$T = 1.25$	Clonic	4	4	0	0	5	3	0	0	−0.47	0.679	1.000
$p_{adj} = 0.404$	Arousal	4	3	1	0	3	0	3	2	1.64	0.061	1.000
	Removal	0	8	0	0	0	7	1	0	1.00	0.167	1.000
	Tonic	8	0	0	0	8	0	0	0	0.00	0.500	1.000
CNS activity	Posture	8	0	0	0	7	0	1	0	1.00	0.167	1.000
$T = 1.28$	Rearing	5	2	1	0	4	2	2	0	0.61	0.277	1.000
$p_{adj} = 0.404$	Palpebral	8	0	0	0	8	0	0	0	0.00	0.500	1.000
Neuromuscular	Gait	8	0	0	0	3	5	0	0	3.42	0.002	0.052
$T = 3.36$	Foot splay	6	1	1	0	6	1	1	0	0.00	0.500	1.000
$p_{adj} = 0.034$	Forelimb	5	2	1	0	2	1	0	5	2.65	0.010	0.221
	Hindlimb	5	3	0	0	0	6	1	1	3.12	0.004	0.090
	Righting	8	0	0	0	5	2	1	0	1.87	0.041	0.824
Physiological	Piloerection	8	0	0	0	8	0	0	0	0.00	0.500	1.000
$T = 1.38$	Weight	6	1	1	0	4	2	0	2	1.17	0.130	1.000
$p_{adj} = 0.404$	Temperature	6	1	1	0	4	3	0	1	0.83	0.210	1.000

is challenging, first because underlying contingency tables for the multivariate categorical responses are very sparse and imbalanced and second because associations of various degrees among the endpoints may mask or enhance effects if not properly taken into account. One simple approach treats the observed scores (and the resulting mean score) as Gaussian, opening up the toolbox for normal-based theory and methods. This might be appropriate if both the number of categories and the sample size are large, which is not the case for the types of multivariate data considered here.

Han et al. (2004) used data from the FOB to illustrate methods for testing a dose-response relationship with multiple correlated binary responses, addressing some of the challenges via an exact, conditional analysis. They transformed the ordinal responses into binary ones (absence or presence of an adverse effect), thereby losing information on severity, and assumed equal correlation among endpoints within a domain coupled with independence of endpoints across different domains. The correlated binary endpoints within a domain were modeled using the distribution of Molenberghs and Ryan (1999), conditioning out two-way interactions and setting higher ones equal to 0. However, associations exist to various degrees both within and across domains, partly because the grouping is often rather subjective and neurotoxic effects defy a simple categorization (Baird et al., 1997). Finally, they assumed a common trend (across the five exposure levels) in the marginal probabilities for all endpoints within a domain and used the significance of this trend to assess potential domain effects.

Klingenberg et al. (2008) treated the data as multivariate ordinal with general correlation structure that they don't model explicitly, by focusing on stochastic order and marginal inhomogeneity among the response vectors as an expression or manifestation of a dose effect, rather than explicitly modeling marginal probabilities under overly simplistic assumptions on both the marginal and joint distributions. All inferences were based on a permutation approach that naturally handles the associations in the multiple endpoints and provides exact significance levels. Now we present a theorem showing that the permutation approach (which assumes exchangeability of entire profiles) is valid for testing marginal homogeneity under a prior assumption of stochastic order. This assumption is usually made implicitly in the multivariate normal case, where one considers a shift in the marginal means but assumes identical covariance matrices, which leads to stochastic order (Müller, 2001). For instance, in the literature on multiple binary endpoints, Westfall and Young (1989), Bilder et al. (2000), and Troendle (2005), all using resampling procedures, assume identical joint distributions under the null, although the hypothesis of interest focuses solely on the margins, as discussed in Agresti and Liu (1999).

The counts displayed in Table 3.2 refer to the $k = 25$ one-way marginal distributions $\{\pi_{ih}(j_h) = \Pr(Y_{ih} = j_h), j_h = 1, \ldots, 4\}$ at the two dose levels $i = 1, 2$. These one-way margins are usually the parameters of interest when it comes to establishing a dose effect. Hence, instead of testing the much narrower H_0, we consider the less restrictive null hypothesis of equality of the vectors of marginal multinomial parameters under the two dose levels. That is, for each adverse event h and outcome category j_h, $\pi_{1h}(j_h) = \pi_{2h}(j_h)$, and we have the hypothesis

$$H_0^{\mathrm{marg}} : \bigcap_{h=1}^{k} \{H_{0h}\} = \bigcap_{h=1}^{k} \left\{ Y_{1h} \overset{d}{=} Y_{2h} \right\}, \qquad (3.4)$$

noting that $H_0 \Rightarrow H_0^{\mathrm{marg}}$. We refer to (3.4) as *simultaneous marginal homogeneity* (SMH, Agresti and Klingenberg, 2005) of the two multivariate ordinal distributions.

Theorem 3.4. *Under the prior assumption* $\boldsymbol{Y}_2 \overset{st}{\geq} \boldsymbol{Y}_1$,

$$H_0 : \boldsymbol{Y}_1 \overset{d}{=} \boldsymbol{Y}_2 \Leftrightarrow H_0^{\mathrm{marg}} : \bigcap_{h=1}^{k} \left\{ Y_{1h} \overset{d}{=} Y_{2h} \right\},$$

$$H_1 : \boldsymbol{Y}_2 \overset{st}{\gneq} \boldsymbol{Y}_1 \Leftrightarrow H_1^{\mathrm{marg}} : \bigcup_{h=1}^{k} \left\{ Y_{2h} \overset{st}{\gneq} Y_{1h} \right\}.$$

Proof. The $\Rightarrow$ direction is trivially true. The reverse implication can be proven via induction. Assume $k = 2$ and SMH. $\boldsymbol{Y}_2 \overset{st}{\geq} \boldsymbol{Y}_1$ implies both $\Pr(Y_{11} \leq j_1, Y_{12} \leq j_2) \geq \Pr(Y_{21} \leq j_1, Y_{22} \leq j_2)$ and $\Pr(Y_{11} > j_1, Y_{12} > j_2) \leq \Pr(Y_{21} > j_1, Y_{22} > j_2)$. Moreover, we have $\Pr(Y_{i1} \leq j_1, Y_{i2} \leq j_2) = 1 - \Pr(Y_{i1} > j_1) - \Pr(Y_{i1} > j_2) + \Pr(Y_{i1} > j_1, Y_{i2} > j_2)$, $i = 1, 2$. It follows that $\Pr(Y_{11} \leq j_1, Y_{12} \leq j_2) - \Pr(Y_{21} \leq j_1, Y_{22} \leq j_2) = \Pr(Y_{11} > j_1, Y_{12} > j_2) - \Pr(Y_{11} > j_1, Y_{12} > j_2)$ for all possible response sequences (j_1, j_2). The left-hand side is nonnegative, while the right-hand side is nonpositive; hence $\boldsymbol{Y}_1 \overset{d}{=} \boldsymbol{Y}_2$ must hold. Let the result be true for $k - 1$. Using Silvester's formula (inclusion-exclusion identity)

$$\Pr(Y_{i1} \leq j_1, \ldots, Y_{ik} \leq j_k) = 1 - \sum_h \Pr(Y_{ih} > j_h) + \sum_{h \neq l} \Pr(Y_{ih} > j_h, Y_{il} > j_l)$$
$$- \ldots + \Pr(Y_{i1} > j_1, \ldots, Y_{ik} > j_k)$$

if k is even or

$$\Pr(Y_{i1} > j_1, \ldots, Y_{ik} > j_k) = 1 - \sum_h \Pr(Y_{ih} \leq j_h) + \sum_{h \neq l} \Pr(Y_{ih} \leq j_h, Y_{il} \leq j_l)$$
$$- \ldots - \Pr(Y_{i1} \leq j_1, \ldots, Y_{ik} \leq j_k)$$

if k is odd, it follows that

$$\Pr(Y_{11} \leq j_1, \ldots, Y_{1k} \leq j_k) - \Pr(Y_{21} \leq j_1, \ldots, Y_{2k} \leq j_k) =$$
$$\Pr(Y_{11} > j_1, \ldots, Y_{1k} > j_k) - \Pr(Y_{21} > j_1, \ldots, Y_{2k} > j_k),$$

as all lower-order marginal and joint probabilities are equal by the induction assumption and hence cancel. Since $\boldsymbol{Y}_2 \overset{st}{\geq} \boldsymbol{Y}_1$, the left-hand side is nonnegative, while the right-hand side is nonpositive, which leaves as the only option $\boldsymbol{Y}_1 \overset{d}{=} \boldsymbol{Y}_2$. Finally, it is easy to see that under $\boldsymbol{Y}_2 \overset{st}{\geq} \boldsymbol{Y}_1$ and not H_0, $H_1 \Leftrightarrow H_1'$.

It follows that, under the prior assumption, testing SMH against the alternative that there is cumulative marginal inhomogeneity $\Pr(Y_{2h} \geq j_h) \geq \Pr(Y_{1h} \geq j_h)$ (with strict inequality for at least one h and j_h) is equivalent to testing equality of the two multivariate distributions against the one-sided alternative that the treatment 2 distribution is stochastically larger and not equal to the treatment 1 distribution. Besides being of interest in its own right, the theorem is important for the validity of a permutation test of SMH and the multiplicity adjustments introduced later. The effect of the prior assumption is to restrict the total parameter space

$$\Omega = \left\{ \pi_i(j_1,\ldots,j_k), i=1,2: \pi_i(j_1,\ldots,j_k) \geq 0, \sum_{(j_1,\ldots,j_k)} \pi_i(j_1,\ldots,j_k) = 1 \right\}$$

spanned by the two unconstrained multinomials to the subset

$$\Omega_r = \{ \pi_i(j_1,\ldots,j_k), i=1,2 : \boldsymbol{Y}_2 \overset{st}{\geq} \boldsymbol{Y}_1 \}.$$

Under Ω_r, rejection of H_0 (or, equivalently, H_0^{marg}) directly leads to H_1 (or, equivalently, cumulative marginal inhomogeneity). This restriction is similar to methods in constrained inference in the univariate case, where one tries to construct more efficient tests under some order restrictions on the parameter space (such as a monotone dose-response) that seem plausible for the given context. A motivation for restricting the parameter space in toxicity studies is based on the observation that it is not unrealistic to expect that an increase in exposure to the toxin results in a shift toward higher outcome categories for some endpoints, while others are unaffected. On the other hand, a shift toward lower categories is unrealistic, as all endpoints describe adverse effects. Hence, we can assume a priori that $\boldsymbol{Y}_2 \overset{st}{\geq} \boldsymbol{Y}_1$, which implies $Y_{2h} \overset{st}{\geq} Y_{1h}$ for all h. Later we present the alternative approach of testing SMH when the prior assumption is not plausible but show that a permutation approach is still reasonable.

3.2 Standardized Test Statistics

Many different statistics can be formed from the vector of marginal sample proportions $\hat{\boldsymbol{\pi}}_i = (\hat{\pi}_{i1}(1), \hat{\pi}_{i1}(2), \ldots, \hat{\pi}_{i1}(c_1), \hat{\pi}_{i2}(1), \ldots, \hat{\pi}_{ik}(c_k))^t$, the most basic one being the difference in marginal sample proportions, $\boldsymbol{d} = \hat{\boldsymbol{\pi}}_2 - \hat{\boldsymbol{\pi}}_1$. With standard results from the underlying multivariate distribution, $\mathrm{E}[\boldsymbol{d}] = \boldsymbol{\pi}_2 - \boldsymbol{\pi}_1$ and $\mathrm{Cov}(\boldsymbol{d})$ is constructed from the variances and covariances within and between endpoints given by

$$\text{Var}[d_h(j)] = \text{Var}[\hat{\pi}_{2h}(j) - \hat{\pi}_{1h}(j)] = \sum_{i=1}^{2} \pi_{ih}(j)[1 - \pi_{ih}(j)]/n_{ih},$$

$$\text{Cov}[d_h(j), d_{h'}(j')] = \sum_{i=1}^{2} [\delta_{hh'}\pi_{ihh'}(j,j') - \pi_{ih}(j)\pi_{ih'}(j')]/n_{ih},$$

where $\delta_{hh'} = 0$ if $h = h'$, $\delta_{hh'} = 1$ if $h \neq h'$, and $\pi_{ihh'}(j,j') = \text{Pr}(Y_{ih} = j, Y_{ih'} = j')$ is the two-way marginal distribution for endpoints h and h'. However, this construction doesn't take advantage of the ordinal nature, and the data are usually too sparse or imbalanced to give a positive definite estimate of $\text{Cov}(\boldsymbol{d})$. Alternatively, let $A = \text{diag}(\boldsymbol{\nu}_h^t, h = 1, \ldots, k)$ be a block-diagonal matrix with scores $\boldsymbol{\nu}_h^t = (\nu_h(1), \nu_h(2), \ldots, \nu_h(c_h))$ as blocks, where $\nu_h(.)$ is some monotone increasing scoring function for the c_h categories of endpoint h. Then, $\boldsymbol{s} = A\hat{\boldsymbol{\pi}}_2 - A\hat{\boldsymbol{\pi}}_1 = A\boldsymbol{d}$ compares mean scores among the two treatments, with covariance matrix $\Sigma = A\text{Cov}(\boldsymbol{d})A^t$. Again, due to sparseness and/or imbalance, Σ and in particular its off-diagonal elements involving the two-way marginals may be impossible to estimate without further simplifying assumptions that lead to a positive definite matrix. One can assume homogeneity across all possible pairs of endpoints for the two-way margins, or, as in Han et al. (2004), equal correlation among endpoints within a domain and no higher-order associations, paired with independence across domains. If neither of these assumptions are plausible, one can always form a test statistic ignoring the covariances between outcomes for different adverse effects, assuming working independence among endpoints.

In any case, it is more efficient to estimate elements of Σ assuming SMH, where one can pool data from the two dose groups to obtain score-type statistics such as $W_0 = \boldsymbol{1}^t\widehat{\Sigma}_0^{-1/2}\boldsymbol{s}/k$, the average of weighted (by the elements of $\widehat{\Sigma}_0^{-1/2}$) mean score differences. Here $\widehat{\Sigma}_0$ is the estimated covariance matrix of $\boldsymbol{s}$ under the SMH hypothesis, where for each $\pi_{ih}(j)$ appearing in $\text{Cov}(\boldsymbol{d})$ the pooled estimator $\hat{\pi}_{+h}(j) = [n_{1h}(j) + n_{2h}(j)]/(n_1 + n_2)$ is plugged in.

Even after pooling the data, estimating off-diagonal elements of Σ can be problematic due to imbalance and sparseness. In general, there are as many as $k(k-1)(c_hc_{h'}-1)$ two-way margins $\pi_{+hh'}(j,j')$ to estimate from the pooled data, with potentially many combinations (j,j') sparsely or never observed for a given endpoint pair (h, h'). By assuming equal correlations in estimating these two-way margins (i.e., $\pi_{+hh'}(j,j')$) are identical for all $k(k-1)$ pairs (h, h')), constructing $\widehat{\Sigma}_0$ and hence W_0 may be possible for data that are not too sparse. Alternatively, as mentioned at the end of the last paragraph, one can consider $W_0' = \boldsymbol{1}^t\Delta_0^{-1/2}\boldsymbol{s}/k$, which, with $\Delta_0 = \text{diag}(\widehat{\Sigma}_0)$, only incorporates as weights the variances and covariances among the c_h categories at a given endpoint and ignores correlations among different endpoints. W_0' is then the average of standardized mean score differences formed at each endpoint.

Without an intrinsic metric, it is common to explicitly or implicitly assign scores to the c_h categories of an ordinal variable to form a test statistic.

Typically used scoring systems $\boldsymbol{\nu}_h$ are based on equally spaced scores, used above, or midranks. If the set of chosen scores is poor in that it badly distorts a numerical scale that underlies the ordered classification, the resulting test will not be sensitive. To address this problem, a statistic with data-driven scoring is given by $T_h^{\max} = \max_{\boldsymbol{\nu}_h}\{z_h(\boldsymbol{\nu}_h)\}$, where the scores $\boldsymbol{\nu}_h^{\max}$ that maximize $z_h(\boldsymbol{\nu}_h)$ are given in Chapter 2. Maximum scores $\boldsymbol{\nu}_h^{\max}$ seem especially appropriate in the toxicity and safety context in the sense that they maximize the contrast (if we were to use normalized scores) of standardized mean score differences between the two doses for the c_h categories of endpoint h.

3.3 Multiple Testing on Endpoints and Domains

Statistics such as W_0' only give a global evaluation of the dose effect and do not indicate which adverse effects (or domains) are responsible for the shift in the marginal distribution. Let $z_h, h = 1, \ldots, k$ denote the k components of $\Delta_0^{-1/2}\boldsymbol{s}$, the standardized mean score differences, which we use as endpoint-specific test statistics. Note that $z_h = [\boldsymbol{\nu}_h^t \text{Cov}_0(\boldsymbol{d}_h)\boldsymbol{\nu}_h]^{-1/2}s_h$ and $W_0' = \frac{1}{k}\sum_{h=1}^k z_h$, where $\boldsymbol{d}_h$ is the vector of differences in marginal sample proportions between the two exposure levels for the c_h categories of endpoint h, $\text{Cov}_0(\boldsymbol{d}_h) = (1/n_1 + 1/n_2)\left[\text{diag}(\hat{\boldsymbol{\pi}}_{+h}) - \hat{\boldsymbol{\pi}}_{+h}\hat{\boldsymbol{\pi}}_{+h}^t\right]$ with $\hat{\boldsymbol{\pi}}_{+h} = (\hat{\pi}_{+h}(1), \ldots, \hat{\pi}_{+h}(c_h))^t$ the pooled sample proportions and $s_h = n\boldsymbol{u}_h^t\boldsymbol{d}_h$. We obtain multiplicity-adjusted endpoint p-values following the closed testing principle of Marcus et al. (1976) that controls the familywise error rate (FWE) in the strong sense. Let H_{0K}^{marg} denote the intersection hypothesis $\bigcap_{h \in K}\{H_{0h}\}$ for a subset $K \subseteq \{1, \ldots, k\}$. Under a full closed testing procedure, there are $2^k - 1$ such intersection hypotheses to test (e.g., more than two million for the FOB data). Hence, we consider the shortcut provided by the step-down approach of Westfall and Young (1993), which is based on the maximum test statistic $\max_h z_h$ and only needs to consider k intersection hypotheses. Under our theorem, which ensures exchangeability, the permutation test for each of these intersection hypothesis is a valid (because it is exact) α-level test, so the FWEs of the stepwise adjustments are guaranteed by the closed testing principle.

However, we will see that the support of the permutation distribution of $\max_h z_h$ is rather discrete (e.g., only 49 different realized values for the FOB data), making it automatically more conservative. This effect is compounded if one considers a test statistic such as $\max_h s_h$ without standardizing by the estimated variance under SMH (e.g., a support of only 18 different values for the FOB data). Note that standardizing is imperative if endpoints are measured on different scales. Because of discreteness, we actually use the mid-p-value approach throughout in finding endpoint (and domain) adjusted p-values. The mid-p-value (Hiriji, 2006) is defined as half the probability (estimated as the percentage over all permutations) of observing a test statistic exactly equal to the one observed plus the probability of observing any larger one. Although strictly controlling the FWE cannot be guaranteed with the mid-p-value approach, simulations presented later show that the distribution of the mid-p-value is very close to uniform under SMH.

The focus in the FOB analysis is not only on the individual adverse effects but also on which of the domains show increased toxicity, testing

$$H_0 : \bigcap_{\text{dom}} \{H_0^{\text{dom}}\} \quad \text{vs.} \quad H_1 : \bigcup_{\text{dom}} \{H_1^{\text{dom}}\},$$

where the individual domain hypotheses $H_0^{\text{dom}} : \bigcap_{h \in \text{dom}} \{H_{0h}\}$ and $H_1^{\text{dom}} :$ $\bigcup_{h \in \text{dom}} \{H_{1h}\}$ are the intersection hypothesis of SMH and its complement over all endpoints within a domain (i.e., the intersection hypothesis H_{0K}^{marg} corresponding to the subset K of endpoints $\{1, \ldots, k\}$ that make up the domain). By grouping endpoints into domains (or body functions for drug safety data), the analysis of domains may increase efficiency for identifying the nature of toxicity effects. This is because some of the endpoints may measure similar effects and thus be redundant. Then, the use of step-down multiplicity corrections at the endpoint level may be conservative and have low statistical power, leading to false negatives, an undesirable feature in toxicology.

By the closed testing principle, the adjusted p-value p_h^{adj} for endpoint h is the maximum over p-values formed by considering all possible intersection hypotheses that include endpoint h. One of these intersections must consist of all the other endpoints in h's domain, which leads to the following fact: The adjusted p-value for a domain is always less than or equal to the minimum of the individual adjusted p-values within the domain; i.e. $p_{\text{Dom}}^{\text{adj}} \leq \min_{h \in \text{Dom}} p_h^{\text{adj}}$. This property results in more power to detect toxicity, albeit only at the domain level. The inequality shows that if a domain test is insignificant, no endpoint within the domain will achieve significance. Conversely, a single significant endpoint within a domain implies a significant domain effect. Finally, if the inequality is strict, domain significance can occur without any single endpoint being significant.

Let $\max_{h \in \text{Dom}_m} z_h$ be the test statistic for the mth domain, $m = 1, \ldots, M$. Then, the step-down adjusted endpoint p-values from the previous section provide an upper bound for the domain tests: A single significant endpoint within a domain implies a significant domain effect. However, taking the maximum focuses only on the strongest toxicity effect, which may not be desirable when one wants to capture effects that accumulate over endpoints within a domain. Then, an appropriate dissonant test statistic is W_0' calculated for each domain m (i.e., $\sum_{h \in \text{Dom}_m} z_h / |\text{Dom}_m|$, where $|.|$ denotes the number of endpoints in the domain). The decision on which type of domain test to use is not straightforward. As a guideline, if one expects endpoints within a domain to be correlated and with similar but possibly moderate effects, a dissonant test statistic is more appropriate than taking the maximum over endpoints within a domain, although multiplicity adjustments at the endpoint level are only possible via the distribution based on the maximum. On the other hand, if one wants to ensure that nonsignificant endpoints are not influencing domain results (perhaps because the domain includes a few nonsignificant endpoints), taking the maximum as the domain test ensures this. Since grouping endpoints into domains (or body functions) is often controversial or not well defined, this robustness feature may be a desirable property.

3.4 Analysis of the FOB Data

We first present some simplistic results based on assigning equally spaced scores $\nu_h(j) = j$ to the categories and treating them as Gaussian. The standard way to deal with multiple endpoints in this context was presented by O'Brien (1984), who used as the statistic the standardized sum of the individual t statistics over all endpoints. Logan and Tamhane (2004) approximate the distribution of this so-called OLS statistic by a t distribution with $(n_1 + n_2 - 2)(1 + 1/k^2)/2$ degrees of freedom. Table 3.2 shows O'Brien's OLS statistic and simple Bonferroni-Holm adjusted p-values based on this approximation for testing an exposure effect in each of the six domains. The table also shows raw and Bonferroni-Holm adjusted p-values for individual endpoints based on forming, for each endpoint, a regular t statistic comparing the mean scores between the two exposures. The domain test indicates a significant shift in severity scores for endpoint(s) in the neuromuscular domain, mostly due to the differences observed for the endpoint "gait".

With regard to a more appropriate permutation analysis, there are $\binom{16}{8} = 12,870$ distinct permutations of the 16 observed animal profiles into the two treatment groups, but many of them lead to identical values of a test statistic such as W_0 or W_0'. For the four endpoints "salivation", "tonic", "palpebral" and "piloerection", the exact same severity categories were observed for all eight animals under both doses, and hence $s_h = 0$ with $\mathrm{Cov}_0(d_h) = \mathbf{0}$. We define $z_h \equiv 0$ for such cases. Alternatively, these four endpoints could have been excluded from an analysis, as they hold no information about marginal inhomogeneity (results under both approaches are almost identical).

We choose $W_0' = \mathbf{1}^t \Delta_0^{-1/2} s/k$ as the test statistic, as with only eight animals per dose group and $k = 25$, it is impossible to obtain a reliable (positive definite) estimate of the full covariance matrix Σ. Note that Δ_0 only needs to be computed once since it is based on the pooled data and is therefore invariant under permutations of profiles. The first panel in Figure 3.1 shows the exact (i.e., using complete enumeration) permutation distribution of W_0'. The 95th percentile of this distribution equals 0.55, and hence the observed W_0' of 1.06 is significant (permutation p-value 0.0002; only 2 of the 12,870 permutations yielded a larger W_0'), indicating a shift in cumulative marginal toxicity for at least one adverse event. For large sample sizes (which is not the case here) and $\pi_{ih}(j_h)$ bounded away from 0 and 1, the standardized mean score differences z_h follow an asymptotic $N(0, 1)$ distribution. Under independence among endpoints, their average (which is W_0') is then asymptotically $N(0, 1/\sqrt{k})$. This distribution is superimposed in Figure 3.1 and one clearly sees that its tails are too light, owing to small sample size and associations among endpoints.

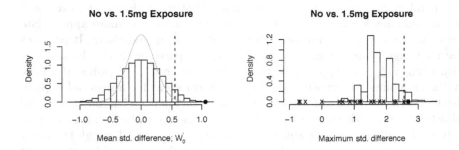

Fig. 3.1. Permutation distributions of global test statistic W_0' (first panel) and $\max_h z_h$ (second panel) for comparing the 1.5 g/kg exposure to the control. Dashed lines indicate 95th percentiles and full circles observed values of test statistics. Crosses indicate observed values for individual endpoints. Grey curves show asymptotic distributions assuming independence among endpoints.

After establishing a global exposure effect, naturally we are interested in which endpoints (or domains) contribute to the significant difference. The raw (i.e., unadjusted) permutation mid-p-value for endpoint h is simply the proportion of permutations that yield a z_h larger than the one observed plus half the proportion of permutations that yield a z_h equal to the one observed. Table 3.3 displays these under all possible permutations. The second panel in Figure 3.1 shows the exact permutation distribution of $\max_h z_h$ and marks the observed z_h's as crosses on the x-axis. This is the starting point for finding the multiplicity adjusted p-values for all endpoints in a step-down procedure, where the permutation distribution of $\max_h z_h$ is used to find the adjusted p-value for the endpoint with the largest z_h statistic (e.g., "gait"). Successive steps delete the endpoint corresponding to $\max_h z_h$ and find the permutation distribution of the maximum over the remaining endpoints, yielding the adjusted p-value for the next largest z_h, "hindlimb" in our example. Table 3.3 lists all resulting step-down adjusted p-values for the $k = 25$ adverse effect in the FOB data. We observe two points. First, even when using $\max_h z_h$ as the global test statistic (instead of W_0', which incorporates *all* standardized differences), we still obtain a significant result, indicating a shift in cumulative marginal toxicity. Second, we can identify the couple of endpoints that are largely responsible for this shift by inspecting the individual multiplicity-adjusted p-values.

The neuromuscular domain is the only one that includes significant adjusted p-values, hence, with the consonant test statistic $\max_h z_h$, it is the only domain that shows increased toxicity at the 1.5 g/kg exposure when compared with a control. However, the domain-adjusted p-values displayed in Table 3.3 are based on the dissonant domain statistic $\sum_{h \in \mathrm{Dom_m}} z_h / |\mathrm{Dom_m}|$ using the full closed testing procedure for the multiplicity adjustments, which now only comprise testing $2^6 - 1 = 63$ intersection hypotheses, a more manageable number than the more than two million that would have been necessary at the endpoint level. We view this dissonant statistic as more appropriate when the focus is on accumulated toxicity over many endpoints. It provides one more significant domain (sensorimotor) compared with the domain test with $\max_{h \in \mathrm{Dom_m}} z_h$. Apparently, the evidence against SMH provided jointly by the endpoints "approach" and "touch" is sufficient for an overall domain effect. The second part of Table 3.3 shows individual and domain raw and adjusted p-values with scores $\boldsymbol{\nu}_h^{\max}$ using the same multiplicity adjusting procedures. Results are comparable to equally spaced scores, although the order of significance of endpoints in the neuromuscular domain is different and we gain one more significant endpoint ("forelimb") at a 5% FWE level. We also tried statistics that avoid assigning scores altogether, such as the likelihood ratio test computed under order restriction or the direct chi-squared test. With these, results for the FOB data were somewhat less significant, and we do not consider them further here. We also do not present results based on using the minimum p-value instead of the maximum test statistic for the p-value adjustments other than saying that they lead to comparable conclusions.

3.5 Violations of Stochastic Order

This assumption seems plausible in a toxicity study where all endpoints describe adverse effects. However, in clinical safety studies measuring side effects of a drug, it is often the case that several marginal sample proportions are larger under the control (placebo) than under some treatment. If these differences are large enough to rule out sampling variability, this is in direct violation of the prior assumption on stochastic ordering. Then, it may not be appropriate to use the permutation approach to construct the null distribution (of any test statistic) under SMH since exchangeability (i.e., IJD) is a stronger assumption.

We consider using the nonparametric bootstrap to test SMH (a parametric bootstrap is problematic because of the modeling issues mentioned above) when the prior assumption does not seem plausible. There are two options on how to resample to obtain a distribution reflecting SMH: We can draw bootstrap samples from centered scores within each group and form a test statistic (such as Welch's t statistic), or we can center an appropriate statistic (such as z_h) obtained from a bootstrap sample of the original responses within each group. Here we use the latter approach, which is preferred by Pollard and van der Laan (2004) because it has asymptotic FWE control for the endpoint analysis. Also, the first approach will not be sensible when using data-dependent scoring. Let $\boldsymbol{Y}_{11}^*, \ldots, \boldsymbol{Y}_{1n_1}^*$ and $\boldsymbol{Y}_{21}^*, \ldots, \boldsymbol{Y}_{2n_2}^*$ denote boot-

Table 3.3. Raw and multiplicity adjusted p-values for endpoints and domains, comparing the 1.5 g/kg exposure to a control, with the test statistic $z_h(\nu_h)$ based on equally spaced and maximum scores.

Domain / Endpoint	Scores $\nu_h = (1,2,3,4)$				Maximum Scores $\nu_h^{\max}$			
	z_h	raw	adj.[a]	adj.[b]	z_h	raw	adj.[a]	adj.[b]
Autonomic[c]		0.214	0.214	0.216		0.135	0.135	0.184
Lacrimation	1.92	0.050	0.331	0.236	1.92	0.050	0.434	0.219
Salivation	0.00	0.500	0.959	0.966	0.00	0.500	0.960	0.930
Pupil	1.15	0.162	0.844	0.726	1.15	0.162	0.875	0.664
Defecation	−0.67	0.633	0.959	0.966	0.00	0.633	0.960	0.930
Urination	−0.71	0.738	0.959	0.966	0.00	0.630	0.960	0.930
Sensorimotor[c]		0.033	0.033	0.002		0.035	0.035	0.095
Approach	2.25	0.019	0.119	0.067	2.31	0.019	0.138	0.064
Click	0.00	0.500	0.959	0.966	0.00	0.500	0.960	0.930
Tail pinch	0.54	0.321	0.927	0.897	0.54	0.321	0.933	0.885
Touch	1.51	0.117	0.566	0.543	1.51	0.117	0.731	0.494
CNS excitability[c]		0.130	0.130	0.076		0.135	0.135	0.190
Handling	1.03	0.182	0.883	0.788	1.03	0.182	0.921	0.819
Clonic	−0.50	0.671	0.956	0.966	−0.50	0.671	0.960	0.930
Arousal	1.61	0.064	0.456	0.463	2.14	0.047	0.235	0.095
Removal	1.03	0.250	0.883	0.788	1.03	0.250	0.921	0.769
Tonic	0.00	0.500	0.959	0.966	0.00	0.500	0.960	0.930
CNS activity[c]		0.107	0.107	0.037		0.113	0.113	0.348
Posture	1.03	0.250	0.833	0.788	1.03	0.250	0.921	0.769
Rearing	0.64	0.280	0.927	0.897	0.67	0.294	0.933	0.836
Palpepral	0.00	0.500	0.959	0.966	0.00	0.500	0.960	0.930
Neuromuscular[c]		0.001	0.001	0.000		0.001	0.002	0.015
Gait	2.70	0.006	0.025	0.013	2.70	0.006	0.041	0.013
Foot splay	0.00	0.500	0.959	0.966	0.00	0.597	0.960	0.930
Forelimb	2.31	0.012	0.096	0.061	2.70	0.009	0.041	0.013
Hindlimb	2.56	0.003	0.044	0.036	2.83	0.003	0.028	0.010
Righting	1.79	0.050	0.352	0.280	1.92	0.050	0.434	0.219
Physiological[c]		0.089	0.089	0.045		0.119	0.119	0.191
Piloerection	0.00	0.500	0.959	0.966	0.00	0.500	0.960	0.930
Weight	1.20	0.133	0.834	0.705	1.55	0.152	0.543	0.383
Temperature	0.87	0.224	0.883	0.843	1.26	0.285	0.875	0.658

[a] Multiplicity adjustments based on the step-down procedure with the maximum test statistic under permutation resampling.
[b] Multiplicity adjustments based on the step-down procedure with the maximum test statistic under bootstrap resampling.
[c] Domain p-values are based on $\sum_{h \in \mathrm{Dom}} z_h / |\mathrm{Dom}|$, with multiplicity adjustments via full closed testing.

strap samples within each group. Under SMH, $E[d_h] = 0$ and we expect the mean score difference s_h^* or its standardized version z_h^* computed from the bootstrap sample to (asymptotically) equal 0. To reflect this, we center s_h^* or z_h^* by subtracting the observed s_h^{obs} or z_h^{obs}, respectively, the estimates of their means under the true data-generating mechanism (which does not necessarily obey SMH).

That is, for each resample we compute $t_h^{1*} = [\boldsymbol{\nu}_h^t \text{Cov}_0(\boldsymbol{d}_h^*)\boldsymbol{\nu}_h]^{-1/2}(s_h^* - s_h^{\text{obs}})$ or $t_h^{2*} = z_h^* - z_h^{\text{obs}}$ with $z_h^* = [\boldsymbol{\nu}_h^t \text{Cov}_0(\boldsymbol{d}_h^*)\boldsymbol{\nu}_h]^{-1/2}s_h^*$. Note that under SMH but not IJD, we can still use pooling to estimate the common marginal probabilities appearing in $\text{Cov}_0(\boldsymbol{d}_h^*)$. We mention both possibilities of first centering s_h^* and then standardizing it or centering z_h^* directly because they show rather different small sample behavior. This is due to the fact that with, for example only eight observations per group and many of them making the same response as in the FOB study, the bootstrap estimate of $\text{Var}(s_h)$ can be very small, leading to many large values of t_h^{1*}. Hence, for the FOB study, raw and adjusted p-values are rather large when using t_h^{1*}, with no endpoint reaching significance even at a 20% FWER level and adjusted domain p-values given by (0.262, 0.011, 0.113, 0.097, 0.000, 0.068). By contrast, step-down adjusted p-values using the bootstrap distribution with t_h^{2*} for endpoints and domains (using $\sum_{h \in \text{Dom}_m} t_h^{2*}$ for the mth domain) are displayed in Table 3.3. Results are comparable to the permutation analysis; however, adjusted p-values are somewhat smaller for the significant endpoints and some of the domains. This is partly due to the fact that the bootstrap distribution of $\max_h t_h^{2*}$ is far less discrete (699 distinct points in 10,000 resamples for the FOB data), but also, as simulations show below, that it doesn't preserve the FWE for small samples in our settings. The bootstrap test for endpoint h is only asymptotically level α (and not exact as with the permutation approach), and with only eight observations per group, we cannot expect to estimate the empirical distribution well. Interestingly, both centering methods give results that are comparable (and almost identical to the permutation approach) when based on maximum scores $\boldsymbol{\nu}_h^{\max}$. The adjusted p-values using t_h^{2*} with maximum scores are also displayed in Table 3.3. Note that with data-dependent scoring, it is necessary to recalculate s_h^{obs} or z_h^{obs} for each bootstrap iteration, as the scoring changes from one to the other.

To evaluate and compare the behavior of the proposed tests, we simulated from a multivariate ordinal distribution with nine endpoints, each with four categories. To simulate under the assumption of SMH without IJD, we generated, for each group, a random vector of 4^9 multinomial probabilities $\{\pi(j_1, \ldots, j_9)\}$ and then used iterative proportional fitting (Deming and Stephan, 1940) to modify these vectors such that they agree in their nine margins. These margins were forced to equal the pooled proportions $\hat{\pi}_{+h}(j), j = 1, \ldots, 4$ for the $h = 1, \ldots, 9$ endpoints of the neuromuscular and sensimotor domains of our FOB study. Two thousand data ets were generated for each group using these multinomial vectors under various combinations of sample sizes (n_1, n_2) under this scenario of SMH without IJD and under the

IJD scenario. Table 3.4 was the true FWE with regard to erroneously (at a nominal 5% level) declaring marginal inhomogeneity with both ways of centering under the bootstrap approach and under the permutation analysis for both the global and individual endpoint analyses and under both assumptions on the true distribution, SMH without IJD and with IJD.

Table 3.4. Part A: Actual FWE (in %) with bootstrap and permutation resampling under SMH but not IJD and under IJD with nine endpoints. The column labeled "global" refers to the proportion out of 2000 generated data sets for which the global test for SMH yielded a p-value less than 5%, while the column labeled "indiv." refers to the proportion out of the same 2000 generated data sets for which at least one of the individual step-down adjusted p-values was less than 5%. Part B: Power (in %) for establishing marginal inhomogeneity when two (gait and approach) out of nine endpoints show marginal inhomogeneity using the global p-value or the step-down adjusted p-value for these two endpoints computed under bootstrap or permutation resampling. Simulation margin of error: $\pm 1\%$.

Part A: FWER control							
	SMH w/o IJD				IJD		
	Bootstrap		Permutation		Bootstrap		Permutation	
(n_1, n_2)	global	indiv.	global	indiv.	global	indiv.	global	indiv.
(8,8)	8.0/9.0	3.9/6.4	5.1	4.3	8.5/9.3	3.4/6.0	5.3	4.7
(25,25)	5.9/6.3	3.9/5.3	5.5	4.6	6.0/6.0	4.3/6.3	5.1	5.2
(50,50)	6.3/5.9	4.8/5.5	5.5	4.3	6.7/6.9	5.5/6.8	5.9	5.7
(25,50)	6.3/6.1	4.4/5.4	5.2	4.9	6.7/6.7	3.9/4.7	5.2	4.4
(25,100)	5.7/5.7	3.8/4.2	4.2	4.7	7.7/7.5	4.1/4.4	5.5	4.6

Part B: Power						
	Bootstrap			Permutation		
(n_1, n_2)	Global	Gait	Approach	Global	Gait	Approach
(8,8)	50/53	48/61	32/42	42	54	37
(25,25)	81/82	87/86	94/96	81	88	96
(50,50)	98/98	99/99	100/100	97	100	100
(25,50)	87/87	96/96	99/99	86	97	99
(25,100)	93/93	98/97	100/100	99	100	100

For small sample sizes, the bootstrap test is liberal, while the permutation test shows good performance, even when the prior assumption of $Y_2 \overset{st}{\geq} Y_1$ (and hence IJD) is not met. For larger sample sizes, all procedures have actual levels close to the nominal ones. Figure 3.2 shows QQ plots comparing the distribution of mid-p-values from the bootstrap and permutation approaches to the uniform distribution for various sample size combinations. Overall, the QQ plots show a nearly straight line, attesting to the near uniform distribution of mid-p-values generated from the bootstrap or permutation analysis.

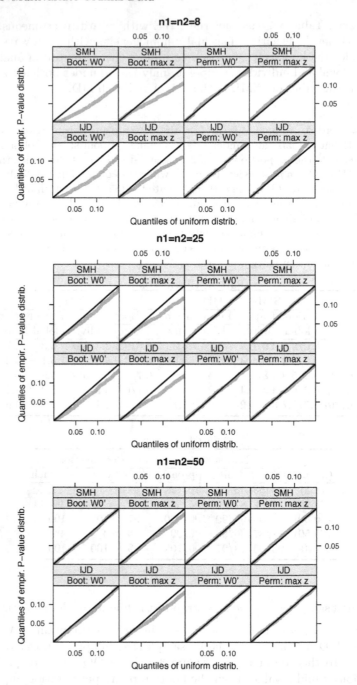

Fig. 3.2. QQ plots of empirical p-value against uniform distribution under the two scenarios SMH without IJD and with IJD. p-values for testing the SMH hypothesis are derived from bootstrapping $T^{1*} = \frac{1}{k} \sum_h t_h^{1*}$ and $T^{2*} = \frac{1}{k} \sum_h t_h^{2*}$ (bootstrap tests 1 and 2, respectively), and the permutation distribution of $W_0' = \frac{1}{k} \sum_h z_h$. The p-value distribution is based on 2000 simulated data sets under each scenario.

In Figure 3.2, we zoom in on the most interesting part up to the 15th percentile. Then, one clearly sees the slight liberal behavior of the bootstrap tests, which vanishes as the sample size increases. The permutation test has excellent performance throughout, even when the prior assumption on stochastic order is not satisfied.

Applications with R functions

All results of the FOB data analysis can be reproduced using R functions available at www.williams.edu/~bklingen. You can directly copy and paste the following into R console:

```
# load data matrices, rows=endpts, cols=animals
> S1 <- read.csv(file="http://lanfiles.williams.edu/~bklingen
+ /SMH/S1.csv",header=FALSE)
#matrix of ordinal observations under control
> S2 <- read.csv(file="http://lanfiles.williams.edu/~bklingen
+ /SMH/S2.csv",header=FALSE)
#matrix of ordinal observations under treatment
> source("http://lanfiles.williams.edu/~bklingen/SMH/SMH.R")
#scource necessary functions
```

Input parameters for **permzh** and **bootzh**:

- c: number of categories (same for all endpoints);
- enum: should complete enumeration be used (permzh only);
- perms/boots: if no complete enumeration, how many random permutations/bootstrap samples should be used (defaults to 10,000);
- scores: scores assigned to categories appearing in S1,S2 (defaults to equally spaced scores), set equal to "max" for maximum scores;
- dom.index: arrange S1, S2 in such a way that endpoints belonging to a common domain are grouped together. Then, dom.index gives the cutpoints of the index in S1, S2 where each domain ends;
- dom.combi: with which function should statistics from domains be combined (defaults to "mean");

Permutation of FOB data with complete enumeration:

```
> fob.perm
+ <-perm.SMH(S1,S2,c=4,enum=TRUE,dom.index=c(5,9,14,17,22,25))
> fob.perm$sig.meanT
# global test using W_0' as test stat
> fob.perm$adj.P
# adjusted P-values (via maxT step down method) for the 25
# endpoints
> fob.perm$adj.P.domain
# adjusted P-values (via full closed testing) for the domains
```

```
> hist(fob.perm$max.T)
# histogram of max_h z_h
```

If complete enumeration is too computationally demanding, use

```
> fob.perm<-perm.SMH(S1,S2,c=4,perms=10000,
+ dom.index=c(5,9,14,17,22,25))
```

using maximum scores:

```
> fob.perm.maxscores<-perm.SMH(S1,S2,c=4,enum=TRUE,
+ scores="max",dom.index= c(5,9,14,17,22,25))
```

user defined scores:

```
fob.perm.userscores <-
perm.SMH(S1,S2,c=4,enum=TRUE,scores=c(1,3,4,5),
+ dom.index=c(5,9,14,17,22,25))
#bootstrap analysis with centering z_h:
fob.boot <- boot.SMH(S1,S2,c=4,boots=10000,
+ dom.index=c(5,9,14,17,22,25))
fob.boot$sig.meanT
#global test using W0' and centering z_h
fob.boot$adj.P
#adjusted P-values (via maxT step down method) for
+ the 25 endpoints
fob.boot$adj.P.domain
#adjusted P-values (via full closed testing) for the 6 domains
#bootstrap analysis with centering s_h and then standardizing:
fob.boot <- boot.SMH(S1,S2,c=4,boots=10000,
+ dom.index=c(5,9,14,17,22,25), centering=2)
fob.boot$sig.meanT
#global test using W0' and centering s_h
fob.boot$adj.P
#adjusted P-values (via maxT step down method) for the
+ 25 endpoints
fob.boot$adj.P.domain
#adjusted P-values (via full closed testing) for the 6 domains
```

4

Multivariate Continuous Data

4.1 Introduction

Most clinical trials are conducted to compare a treatment group with a control group on multiple endpoints. A classic example is chronic obstructive pulmonary disease (COPD), where there are a number of different types of potential endpoints (Pocock et al., 1987). Lung function tests are usually considered as key endpoints in COPD trials. However there are a large number of alternative lung function measurements that are believed to indicate efficacy in COPD (e.g. peak expiratory flow rate, forced expiratory volume, forced vital capacity, etc.). It therefore seems logical to consider multivariate approaches when assessing if the treatment improves the respiratory function compared with the control.

In principle, the treatment is usually deemed better than the control if all components of its responses are larger. In some practical situations, it may be difficult to show that each component is better. Instead, the treatment will be *superior* if at least one of its response components is greater than that of the control. Hotelling's T^2 test is an obvious choice when interest lies in determining whether treatment is demonstrating an overall nondirectional effect on a number of endpoints. However, this standard approach lacks power because it does not take account of the fact that the direction of a response is known in advance, failing to incorporate the restrictions on the null and alternative parameter spaces.

We focus on the case where all endpoints are primary, and provide a comprehensive review of the vast literature and some new results focusing on the statistical aspects. Beginning with O'Brien's seminal paper (1984), the problem of constructing one-sided tests for comparing multivariate treatment effects has received much attention in the literature. Most of the methods developed have been based on the multivariate normal distribution. In particular, likelihood ratio tests and approximations (Kudo, 1963; Perlman, 1969; Tang et al., 1989; Wang and McDermott, 1998) for one-sided alternatives have been considered. However, these tests have some unappealing properties. Sil-

D. Basso et al., *Permutation Tests for Stochastic Ordering and ANOVA*, Lecture Notes in Statistics, 194, DOI 10.1007/978-0-387-85956-9_4,
© Springer Science+Business Media, LLC 2009

vapulle (1997) pointed out that, when all the observed treatment differences are negative and the endpoints are positively correlated, the test can still reject the null hypothesis in favor of the one-sided alternative (see also Logan, 2003).

Multivariate statistics offer a practical way to combine multiple endpoints into a single test, therefore avoiding issues with multiple testing and the requirement of alpha adjustments. In the form presented it will be difficult to interpret a statistically significant finding in clinical terms. In clinical trials, the statistical and clinical significance of the individual variables remains very important even if global tests or composite tests indicate an overall effect.

Broadly speaking, there are two inferential goals when dealing with multiple endpoints:

i.   to establish an overall treatment effect using a test of the global null hypothesis of no differences on any of the endpoints against a one-sided alternative; and

ii. to identify the individual endpoints on which the treatment is better than the control.

If it is not assumed a priori, another condition that needs to be satisfied by the results from a clinical trial is the requirement of a positive statement on *noninferiority* for all variables; that is,

iii. to show that the treatment is not much worse on any of the endpoints.

Noninferiority means that the effect of the treatment is not worse than that of the control by more than a specified margin. Thus, the treatment will be preferred if it is superior for at least one of the endpoints and not inferior for the remaining endpoints.

4.2 Testing Superiority

Suppose that there are two independent treatment groups with n_1 and n_2 subjects on each of whom $k \geq 2$ endpoints are measured. Let Y_{ihj} denote the hth response of the jth subject in the ith treatment group, and suppose we have a total of $n = n_1 + n_2$ subjects randomly assigned to the two treatments, such that $Y_{11}, \ldots, Y_{1n_1}$ are n_1 i.i.d. observations and, independently, $Y_{21}, \ldots, Y_{2n_2}$ are n_2 i.i.d. observations.

It is desired to test wheter the samples have been generated from the same probability law $H_0 : Y_1 \overset{d}{=} Y_2$ against the alternative that the Y_2 distribution is stochastically larger and not equal to the Y_1 distribution; i.e., $H_1 : Y_1 \overset{st}{\leq} Y_2$. Then, the assumed nonparametric model is $\mathcal{M} = \{Y_1, Y_2 \in \mathbb{R}^k : Y_1 \overset{st}{\leq} Y_2\}$. Under this model, we have shown that testing H_0 against H_1 is equivalent to the union-intersection testing formulation

$$H_0^{\text{st}} : \bigcap_{h=1}^{k} \{Y_{1h} \overset{d}{=} Y_{2h}\} \quad \text{against} \quad H_1^{\text{st}} : \bigcup_{h=1}^{k} \{Y_{1h} \overset{\text{st}}{\lneq} Y_{2h}\}.$$

To fix ideas and to facilitate comparison with previous work, we assume here the two treatments with multivariate normal distributions, and the responses are compared on the basis of their mean response vectors. For treatment group i, assume that $\boldsymbol{Y}_{ij}$, $j = 1, \ldots, n_i$, are i.i.d. random vectors from a k-variate normal distribution with mean vector $\boldsymbol{\mu}_i = (\mu_{i1}, \ldots, \mu_{ik})$ and covariance matrix $\boldsymbol{\Sigma}_i$. In the homoscedastic case, we assume $\boldsymbol{\Sigma}_1 = \boldsymbol{\Sigma}_2 = \boldsymbol{\Sigma}$. The elements of $\boldsymbol{\Sigma}$ are $\sigma_{hh} = \text{Var}(Y_{ih})$ and $\sigma_{hl} = \text{Cov}(Y_{ih}, Y_{il})$, $1 \le h < l \le k$. Denote by $\boldsymbol{R}$ the corresponding correlation matrix with elements $\rho_{hl} = \text{Corr}(Y_{ih}, Y_{il}) = \sigma_{hl} / \sqrt{\sigma_{hh} \sigma_{ll}}$. In the heteroscedastic case, the elements of $\boldsymbol{\Sigma}_i$ and $\boldsymbol{R}_i$ will be denoted by $\sigma_{i,hl}$ and $\rho_{i,hl}$, respectively. Let $\boldsymbol{\theta} = \boldsymbol{\mu}_2 - \boldsymbol{\mu}_1 = (\theta_1, \ldots, \theta_m)$ denote the vector of mean differences.

Under the multivariate normality assumption, $\boldsymbol{Y}_1 \overset{\text{st}}{\le} \boldsymbol{Y}_2$ if and only if $\theta_h \ge 0$ for all $h = 1, \ldots, k$ and $\boldsymbol{\Sigma}_1 = \boldsymbol{\Sigma}_2$ (Müller, 2001); that is, we have to assume homoscedasticity and restrict the parameter space $\Omega = \{\boldsymbol{\theta} : \boldsymbol{\theta} \in \mathbb{R}^k\}$ to the positive orthant at the origin (i.e., $\Omega^+ = \{\boldsymbol{\theta} : \boldsymbol{\theta} \ge 0\}$). For $k = 2$, the regions of the parameter space corresponding to $H_0 : \boldsymbol{\theta} \le 0$ against $H_1 : \boldsymbol{\theta} \gneq 0$ and $H_0^{\text{st}} : \boldsymbol{\theta} = 0$ and $H_1^{\text{st}} : \boldsymbol{\theta} \gneq 0$ are shown in Figure 4.1.

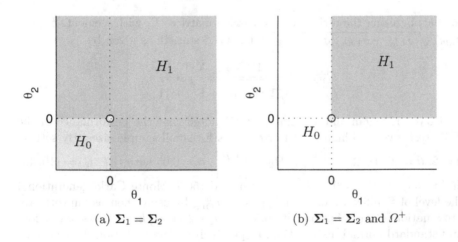

(a) $\boldsymbol{\Sigma}_1 = \boldsymbol{\Sigma}_2$ (b) $\boldsymbol{\Sigma}_1 = \boldsymbol{\Sigma}_2$ and Ω^+

Fig. 4.1. Regions of the parameter space.

For testing each component hypothesis $H_{0h} : \theta_h = 0$ against $H_{1h} : \theta_h > 0$, consider Student's t statistic

$$T_h = \frac{\bar{Y}_{2h} - \bar{Y}_{1h}}{\hat{\sigma}_h \sqrt{\frac{1}{n_1} + \frac{1}{n_2}}},$$

where $\hat{\sigma}_h^2$ is the pooled sample variance. O'Brien (1984) derived the ordinary least squares (OLS) statistic

$$T_{\text{OLS}} = \frac{1^t T}{\sqrt{1^t \hat{R} 1}}$$

and the generalized least squares (GLS) statistic $T_{\text{GLS}} = \frac{1^t \hat{R}^{-1} T}{\sqrt{1^t \hat{R}^{-1} 1}}$, where $\hat{R}$ is the estimated correlation matrix, T is a vector of the t statistics, and 1 is a vector of all 1's. Both the OLS and GLS statistics are standardized weighted sums of the individual t statistics for the k endpoints. However, the linear combination of the t statistics used in the GLS test can have some negative weights, which can lead to violation of the *monotonicity requirement* of rejection regions in the sense that, as the sample treatment differences become more negative, the test statistic can get larger, thereby increasing the chance of rejecting the null hypothesis, being "untenable from a practical viewpoint" (Pocock et al., 1987). A more appealing test would be one that is strictly nondecreasing in the direction of each individual endpoint statistic.

Logan and Tamhane (2004) proposed an extension of O'Brien's OLS test to the heteroscedastic case by standardizing the observations as

$$X_{ihj} = \frac{Y_{ihj}}{\sqrt{\hat{\sigma}_{1,hh} + \hat{\sigma}_{2,hh}}}$$

so that, by using the estimated covariance matrices $\hat{\Gamma}_i$ with elements $\hat{\gamma}_{i,hl} = \hat{\sigma}_{i,hl}/\sqrt{(\hat{\sigma}_{1,hh} + \hat{\sigma}_{1,hh})(\hat{\sigma}_{1,ll} + \hat{\sigma}_{1,ll})}$, the OLS statistic is given by

$$T_{\text{OLS}}^{\text{het}} = \frac{1^t (\bar{X}_2 - \bar{X}_1)}{\sqrt{1^t (\hat{\Gamma}_1/n1 + \hat{\Gamma}_2/n_2) 1}}.$$

However, Logan and Tamhane (2004) noted that the performance of the $T_{\text{OLS}}^{\text{het}}$ test gives too high type I error rates for small sample sizes. By setting $k = 2$, $\theta = (0,0)$, $R_1 = \begin{bmatrix} 1 & 0 \\ 0 & 1 \end{bmatrix}$, $R_2 = \begin{bmatrix} 1 & \rho \\ \rho & 1 \end{bmatrix}$, $\alpha = 0.05$ with $(n_1, n_2) = (10, 40)$ or $(n_1, n_2) = (100, 400)$, we have estimated (5000 Monte Carlo generations) the level of significance of both T_{OLS} and $T_{\text{OLS}}^{\text{het}}$ by using their asymptotic approximations (t-distribution with $0.5(n_1 + n_2 - 2)(1 + 1/k^2)$ degrees of freedom and standard normal distribution, respectively) or permutation distributions (2000 permutations). The resulting empirical cumulative distributions of the p-values obtained are shown in Figure 4.2. Our simulations confirm that the use of the standard normal or resampling-based critical values in performing the $T_{\text{OLS}}^{\text{het}}$ tests give too high type I error rates for small sample sizes.

Lehmacher et al. (1991) point out that Bonferroni and, by extension, global tests based on the maximum test statistics are useful for detecting one highly significant difference among a group of otherwise barely significant or nonsignificant differences, whereas O'Brien's tests, based on the sum, succeed in

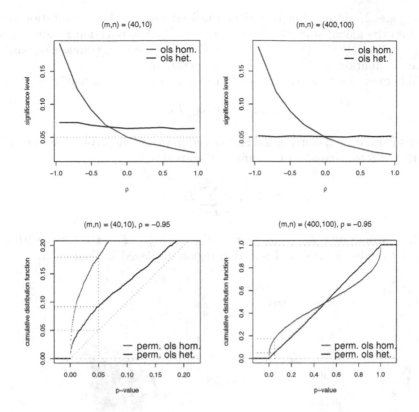

Fig. 4.2. Cumulative distribution functions of the p-values of T_{OLS} and $T_{\mathrm{OLS}}^{\mathrm{het}}$.

rejecting the global null against alternatives closer to the diagonal, meaning a group of similar treatment effects, none of which may achieve significance.

We illustrate the different rejection regions by means of a numeric example, given in Table 4.1.

Table 4.1. A fictitious example of data with bivariate observations from eight subjects per group. The subjects (16 in total) are assumed to be independently and identically distributed according to a multivariate normal distribution with unknown covariance.

Y_{11}	0.845	1.725	0.261	−0.079	0.962	2.408	−0.198	2.021
Y_{12}	−0.035	−0.606	−0.717	1.668	1.529	1.206	−0.141	0.524
Y_{21}	1.022	−0.205	1.334	0.255	−0.165	0.856	−0.480	0.476
Y_{22}	−0.607	0.143	−0.467	−0.641	−1.296	−0.535	−0.477	1.083

In general, given a function ψ of the T statistics, the permutation test based on the global statistic $T_\psi = \psi(T_1, \ldots, T_k)$ has rejection region $R_\psi^* = \{(T_1, \ldots, T_k) : T_\psi > c_\psi^*(1-\alpha)\}$, where $c_\psi^*(1-\alpha)$ is the $1-\alpha$ (random) quantile of the distribution of T_ψ obtained by permutations.

Well-known cases are the max-T test based on the maximum of T statistics

$$T_{\max} = \max_{h=1,\ldots,k} T_h,$$

with its permutation distribution and rejection region displayed in Figure 4.3, and the OLS test based on the sum of T-statistics

$$T_{\text{sum}} = \sum_{h=1}^{k} T_k,$$

which is permutationally equivalent to $T_{\text{sum}} = \sum_{h=1}^{k} T_k / (\mathbf{1}^t \hat{\mathbf{R}} \mathbf{1})^{1/2}$, with its permutation distribution and rejection region displayed in Figure 4.4.

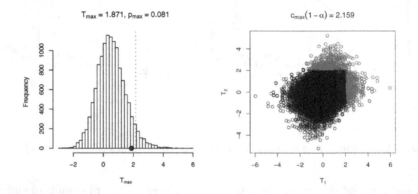

Fig. 4.3. Permutation distribution and rejection region of the $T_{\max}$ statistic.

Essentially, global tests simplify the problem by testing against an alternative that specifies a departure from $\theta = \mathbf{0}$ in a prechosen direction of the positive orthant Ω^+. For example, for testing $(\theta_1, \theta_2) = (0,0)$ against the fixed alternative $(\theta_1', \theta_2') \in \Omega^+$, the most powerful test rejects for large values of

$$T_{\text{mp}} = (\theta_1' - \rho\theta_2')T_1 + (\theta_2' - \rho\theta_1')T_2.$$

However, one of the coefficients of T_h in the optimal test statistic can be negative. Bittman et al. (2006) give an example with $\theta_1' = 2$, $\theta_2' = 8$, and $\rho = 1/2$, in which the optimal test statistic is $-2T_1 + 7T_2$. It is perhaps surprising that this test is not monotone in each T_h. That is, decreasing T_1 increases the value of the test statistic, and the test rejects with probability tending to one as $T_1 \to -\infty$.

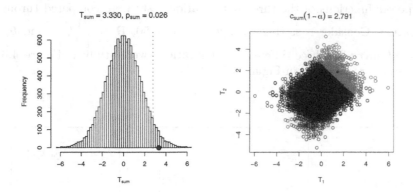

Fig. 4.4. Permutation distribution and rejection region of the T_{sum} statistic.

A test procedure with a more uniform power performance can be obtained by considering a weighted sum of T statistics

$$T_{\text{wsum}} = \sum_{h=1}^{k} w_h T_h,$$

where $\boldsymbol{w} = (w_1, \ldots, w_k)$ is a given direction in the positive orthant, and thus each w_h is nonnegative. By using $w_h = |\hat{\theta}_h|/\hat{\sigma}_h$, we obtain a sum of signed t squared (or F) statistics

$$T_{\text{ss}T^2} = \sum_{h=1}^{k} \text{sign}(T_h) T_h^2,$$

with its permutation distribution and rejection region displayed in Figure 4.5.

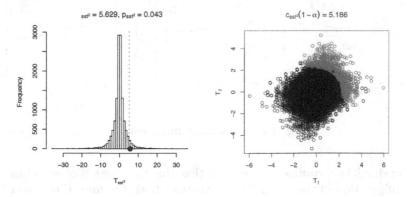

Fig. 4.5. Permutation distribution and rejection region of the $T_{\text{ss}T^2}$ statistic.

The power functions of the three permutation tests were compared through simulation. Consider a setting with $n_1 = n_2 = 50$, $\mathbf{R} = \begin{bmatrix} 1 & \rho \\ \rho & 1 \end{bmatrix}$, and $\boldsymbol{\theta} = r(\cos rad, \sin rad)$. Three different configurations were studied (Figure 4.6), and results are reported in Figure 4.7.

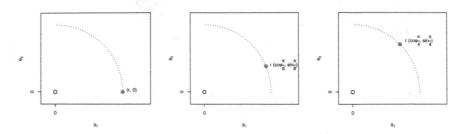

Fig. 4.6. Simulation study configuration.

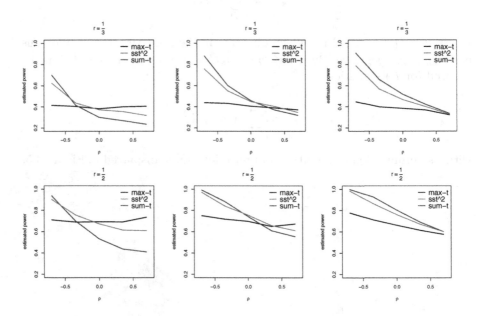

Fig. 4.7. Simulation study results.

The simulation results demonstrate that the T_{ssT^2} test is more robust to the configuration of the mean differences than both the max-T test and the OLS test. It is found to be especially advantageous when the correlations between the endpoints are low or negative. There is very little loss of power in other situations.

A testing method is consonant when the rejection of an intersection hypothesis implies the rejection of at least one of its component hypotheses (Hochberg and Tamhane, 1987). With a dissonant test (such as the sum-T), if the null hypothesis $H_0 : \boldsymbol{\theta} = \mathbf{0}$ is rejected but the statistician cannot reject either of the individual hypotheses, then compelling evidence has not been established to promote a particular drug indication. Lack of consonance makes interpretation awkward.

Bittman et al. (2006) show how to modify a dissonant test into a consonant one with a rejection region of the form $R_{c\psi}^* = \{(T_1, \ldots, T_k) : T_\psi > c_{c\psi}^*(1-\alpha) \cap [T_1 > c_1^*(1-\alpha) \cup T_2 > c_2^*(1-\alpha)]\}$, where $c_h^*(1-\alpha)$ is the $1-\alpha$ quantile of the permutation distribution of the hth T statistic and $c_{c\psi}^*(1-\alpha)$ is determined so that, under $H_0 : \boldsymbol{\theta} = \mathbf{0}$, the region has probability α. Figure 4.8 shows the modified rejection region of the consonant sum-T and ssT^2 tests.

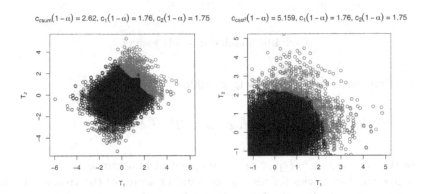

Fig. 4.8. Rejection region of the consonant tests.

The power functions of the max-T, consonant sum-T and ssT^2 were compared through simulation with the same setting used previously. The powers to reject the null hypothesis are reported in Figure 4.9.

4.3 Testing Superiority and Noninferiority

The foregoing testing methods do not suffice for establishing that the treatment will be preferred with respect to the control if it is also required to provide statistical evidence simultaneously to establish that the treatment is not inferior to the control with respect to every variable.

The treatment is regarded as noninferior to the control on the hth endpoint if $\theta_h > -\epsilon_h$ and the hypotheses on noninferiority for all variables have the

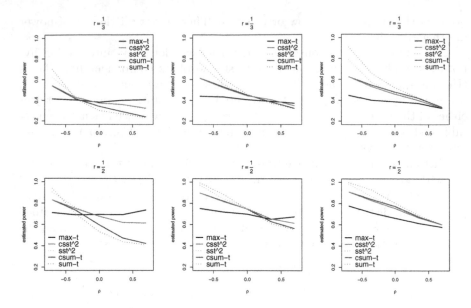

Fig. 4.9. Simulation study results.

intersection-union formulation Berger (1998)

$$H_0^\epsilon : \bigcup_{h=1}^{k} \{\theta_h \le -\epsilon_h\} \quad \text{against} \quad H_1^\epsilon : \bigcap_{h=1}^{k} \{\theta_h > -\epsilon_h\},$$

where the constants $\epsilon = (\epsilon_1, \ldots, \epsilon_k)$ are specified in advance.

Define the t statistics for testing the noninferiority of the treatment with respect to the control on the hth endpoint by

$$T_h^\epsilon = \frac{\bar{Y}_{2h} - \bar{Y}_{1h} + \epsilon_h}{\hat{\sigma}_h \sqrt{\frac{1}{n_1} + \frac{1}{n_2}}} = T_h + \frac{\epsilon_h}{\hat{\sigma}_h \sqrt{\frac{1}{n_1} + \frac{1}{n_2}}} = T_h + \delta_h$$

and by using the intersection-union approach results in the min-T test (Laska and Meisner, 1989) that rejects H_0^ϵ at level α if

$$\min_{1 \le h \le k} T_h^\epsilon > t_{1-\alpha,\nu},$$

where, by assuming normality, $t_{1-\alpha,\nu}$ is the $1-\alpha$ quantile of the t distribution with $\nu = n_1 + n_2 - 2$ degrees of freedom. The resulting rejection region has the form $R^\epsilon = \{(T_1, \ldots, T_k) : (T_1, \ldots, T_k) > (t_{1-\alpha,\nu} - \delta_1, \ldots, t_{1-\alpha,\nu} - \delta_k)\}$.

The overall testing problem becomes an intersection-union combination of intersection-union and union-intersection testing problems,

$$H_0^\circ : H_0 \cup H_0^\epsilon \quad \text{against} \quad H_1^\circ : H_1 \cap H_1^\epsilon.$$

For $k = 2$, the regions of the parameter space corresponding to H_0° and H_1° are shown in Figure 4.10.

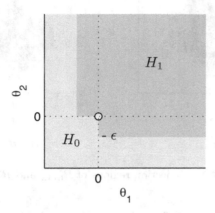

Fig. 4.10. Regions of the parameter space of H_0° and H_1°.

We can proceed through a two-step procedure, where noninferiority must be proven first and superiority has to be shown subsequently by rejecting H_0° if $(T_1, \ldots, T_k) \in R_\psi^* \cap R^\epsilon$. This rejection region is shown in Figure 4.11(a) with $\epsilon = 1/2$ and $R_\psi^* = R_{\max}^*$. Thus $T_{\max} = 1.871 \le 2.16 = c_{\max}^*(1 - \alpha)$ and we don't reject H_0 and hence also H_0°.

However, as noted by Tamhane and Logan (2004), this procedure is conservative because it requires that the type I error probability be controlled separately for H_0^ϵ and H_0. To apply the less conservative intersection-union test of H_0° at level α, we should solve for d to find the $(1 - \alpha)$ critical value of the superiority test such as $\Pr(\min_{1 \le h \le k} T_h^\epsilon > t_{1-\alpha, \nu} \cap T_\psi > d) = \alpha$. We give the following permutation-based algorithm, which determines $R_{tl}^* = \{(T_1, \ldots, T_k) : (T_1, \ldots, T_k) > (t_{1-\alpha, \nu} - \delta_1, \ldots, t_{1-\alpha, \nu} - \delta_k) \cap T_\psi > c_{tl}^*(1 - \alpha)\}$:

(Testing noninferiority and superiority)
step 0   If $\min_{1 \le h \le k} T_h^\epsilon \le t_{1-\alpha, \nu}$, then accept $H_0^\epsilon \subset H_0^\circ$ and stop; otherwise
step 1   compute $(T_1^*(b), \ldots, T_k^*(b))$ from the bth permutation, $b = 1, \ldots, B$,
obtaining c_{tl}^* as the $(1 - \alpha)$ quantile of $\{T_\psi^*(b) : \min_h T_h^{\epsilon *}(b) > t_{1-\alpha, \nu}\}$
step 2   If $T_\psi \le c_{tl}^*(1 - \alpha)$, then accept $H_0 \subset H_0^\circ$, otherwise reject H_0°.

The rejection region R_{tl}^* is shown in Figure 4.11(b) with $\epsilon = 1/2$ and $T_\psi = T_{\max} = 1.871$, so that $c_{tl}^*(1 - \alpha) = 1.6$ compared with $c_{\max}^*(1 - \alpha) = 2.16$.

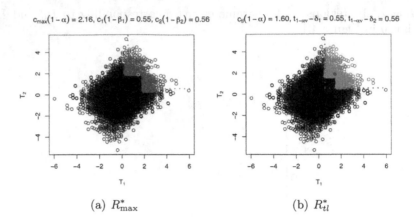

(a) $R^*_{\max}$ (b) R^*_{tl}

Fig. 4.11. Rejection regions of $R^*_{\max}$ and R^*_{tl}.

4.3.1 Applications with R functions

For performing the statistical tests illustrated in the previous sections, load the data from Table 4.1, where X represents class labels and Y the vector of data. Compute the T_h's statistics on observed and permuted data by

```
> source("ptest2s.R")
> set.seed(0)
> B <- 5000
> T <- ptest2s(Y,X,B,"Student")
```

Perform the combined tests $T_{\max}$, T_{sum}, and $T_{\mathrm{ssT^2}}$ by

```
> maxT <- apply(T,1,max)
> sum(maxT[-1]>=maxT[1])/(B-1)
[1] 0.08861772 #p-value sumT
> sumT <- apply(T,1,sum)
> sum(sumT[-1]>=sumT[1])/(B-1)
[1] 0.02680536 #p-value maxT
> ssT2 <- apply(T,1,function(row){sum(sign(row)*(row^2))})
> sum(ssT2[-1]>= ssT2[1])/(B-1)
[1] 0.04620924 #p-value ssT2
```

Check if at least one T_h is greater than $c^*_h(1-\alpha)$

```
> alpha <- 0.05
> c1 <- sort(T[-1,1])[B*(1-alpha)]
> c2 <- sort(T[-1,2])[B*(1-alpha)]
> T[1,] > c(c1,c2)
[1] FALSE  TRUE
```

and compute the p-values of the consonant T_{csum} and T_{cssT^2} tests

```
> ind <- (T[,1]>c1|T[,2]>c2)
> sum(sumT[ind]>=sumT[1])/(B-1)
[1] 0.02580516 #p-value csumT
> sum(ssT2[ind]>=ssT2[1])/(B-1)
[1] 0.04540908 #p-value cssT2
```

Check noninferiority with $\epsilon = 1/2$

```
> eps <- 0.5
> ct <- qt(1-alpha,length(Y)-2)
> delta <- eps/(mean(apply(Y,2,sd)))
> T[1,] + delta > c(ct,ct)
[1] TRUE TRUE
```

but because we don't reject superiority by T_{max}, simultaneously test noninferiority and superiority by

```
> ind2 <- (T[,1]+delta>ct & T[,2]+delta>ct)
> sum(maxT[ind2]>=maxT[1])/(B-1)
[1] 0.01880376 # p-value
```

4.4 Several Samples

In a dose-response experiment, r doses of a treatment are administered to independent groups of n_i subjects, $i = 1, \ldots, r$. Let $\boldsymbol{Y}_{is} = (Y_{1is}, \ldots, Y_{kis}) \in \mathbb{R}^k$ be a vector of response on k variables for the sth subject randomly assigned to treatment dose i, $i = 1, \ldots, r$, $s = 1, \ldots, n_i$. We wish to test the global null hypothesis

$$H_0 : \boldsymbol{Y}_1 \overset{d}{=} \ldots \overset{d}{=} \boldsymbol{Y}_r$$

against the alternative

$$H_1 : \boldsymbol{Y}_1 \overset{st}{\leq} \ldots \overset{st}{\leq} \boldsymbol{Y}_r,$$

with at least one strict inequality. If we assume that $\boldsymbol{Y}_i \sim N_k(\boldsymbol{\mu}_i, \boldsymbol{\Sigma}_i)$, $i = 1, \ldots, r$, then $\boldsymbol{Y}_1 \overset{st}{\leq} \ldots \overset{st}{\leq} \boldsymbol{Y}_r$ if and only if $\mu_{h1} \leq \ldots \leq \mu_{hr}$ for all $h = 1, \ldots, k$ and $\boldsymbol{\Sigma}_1 = \ldots = \boldsymbol{\Sigma}_r$ (Müller, 2001). This result generalizes the fact that if $Y_i \sim N(\mu_i, \sigma_i^2)$, $i = 1, \ldots, r$, then $Y_1 \overset{st}{\leq} \ldots \overset{st}{\leq} Y_r$ if and only if $\mu_1 \leq \ldots \leq \mu_r$ and $\sigma_1^2 = \ldots = \sigma_r^2$. More generally, whenever the random vectors have the same dependence structure (that is, by assuming a location-shift model, $\boldsymbol{Y}_i = \boldsymbol{Y} + \boldsymbol{\delta}_i$, $i = 1, \ldots, r$, with $\boldsymbol{\delta}_1 \leq \ldots \leq \boldsymbol{\delta}_r \in \mathbb{R}^k$), then multivariate stochastic ordering conditions may be provided simply by examining the marginal distribution functions of each of the components of the random vectors. However, by assuming that $\boldsymbol{Y}_1 \overset{st}{\leq} \ldots \overset{st}{\leq} \boldsymbol{Y}_r$ holds, it is straightforward to extend the results given in Theorem 3.4 to the r-sample case. If SMH holds (i.e., $Y_{h1} \overset{d}{=} \ldots \overset{d}{=} Y_{hr}$, $h = 1, \ldots, k$), it follows that testing H_0 against H_1 is equivalent to testing

$$\bigcap_{h=1}^{k} \{H_{0h}\} : \bigcap_{h=1}^{k} \left\{ Y_{h1} \overset{d}{=} \ldots \overset{d}{=} Y_{hr} \right\}$$

against

$$\bigcup_{h=1}^{k} \{H_{1k}\} : \bigcup_{h=1}^{k} \left\{ Y_{h1} \overset{st}{\leq} \ldots \overset{st}{\leq} Y_{hr} \text{ and not } H_{0h} \right\}.$$

Testing H_{0h} against H_{1h} has received considerable attention in the past. Silvapulle and Sen (2005) is the most recent reference using likelihood methodology. Another popular test based on ranks is given in Jonckheere (1954) and Terpstra (1952), and recently ElBarmi and Mukerjee (2005) suggested a stepwise test based on a method of Hogg (1962).

Alternatively, because Y_h regression dependent on X (i.e., $F_{h1}(y) = \Pr\{Y_h \leq y | X = 1\} \geq \ldots \geq \Pr\{Y_h \leq y | X = r\} = F_{hr}(y)$, $\forall y \in I\!R$) implies that the pair (Y_h, X) is positive quadrant dependent (i.e., $F_{h(1:i)}(y) = \Pr\{Y_h \leq y | X \leq i\} \geq \Pr\{Y_h \leq y | X > i\} = F_{h(i+1:r)}(y)$, $\forall y \in I\!R$, $i = 1, \ldots, r - 1$), one can consider testing

$$H_{0h}^{\star} : \bigcap_{i=1}^{r-1} \left\{ F_{h(1:i)}(y) = F_{h(i+1:r)}(y), \quad \forall y \in I\!R \right\}$$

by noting that $H_{0h} \Leftrightarrow H_{0h}^{\star}$ against

$$H_{1h}^{\star} : \bigcup_{i=1}^{r-1} \left\{ F_{h(1:i)}(y) \gneqq F_{h(i+1:r)}(y), \quad \forall y \in I\!R \right\},$$

which is less restrictive than H_{1h} (i.e., $H_{1h} \Rightarrow H_{1h}^{\star}$). However, under the assumption $\boldsymbol{Y}_1 \overset{st}{\leq} \ldots \overset{st}{\leq} \boldsymbol{Y}_r \Rightarrow Y_{h1} \overset{st}{\leq} \ldots \overset{st}{\leq} Y_{hr}$, we have $H_{1h} \Leftrightarrow H_{1h}^{\star}$, $h = 1, \ldots, k$, where for testing

$$H_{0hi} : F_{h(1:i)} = F_{h(i+1:r)} \quad \text{against} \quad H_{1hi} : F_{h(1:i)} \gneqq F_{h(i+1:r)}$$

we pool together the first i groups and the last $(r - i)$ groups, $i = 1, \ldots, r - 1$.

For testing H_{0hi} against H_{1hi}, the permutation test based on the usual two-sample t statistic or equivalently on

$$T_{hi} = \sum_{i=i+1}^{r} \sum_{s=1}^{n_i} Y_{his} \tag{4.1}$$

is the most powerful permutation test against the normal alternatives that common variance among all permutation tests that are unbiased and of level α. Obviously, other two-sample test statistics can be considered, for instance the Kolmogorov-Smirnov statistic

$$KS_{hi} = \max_{y \in \mathbb{R}}\{\hat{F}_{h(i+1:r)}(y) - \hat{F}_{h(1:i)}(y)\}, \tag{4.2}$$

where $\hat{F}_{h(a:b)}(y) = \frac{\sum_{i=a}^{b}\hat{F}_{hi}(y)n_i}{\sum_{i=a}^{b}n_i}$ and $\hat{F}_{hi}$ denotes the usual empirical c.d.f.

Then, we may combine first with respect to pseudogroups (i.e., for testing $H_{0h} : \bigcap_{i=1}^{r-1}\{H_{0hi}\}$), obtaining $T_h = \psi(T_{hi}, i = 1, \ldots, r-1)$, and next with respect to variables (i.e. for testing $H_0 : \bigcap_{h=1}^{k}\{H_{0h}\}$) obtaining the global test statistic $T = \psi(T_h, h = 1, \ldots, k)$.

For testing H_0 against H_1, Dietz (1998) proposed an asymptotically distribution-free test that generalizes the Jonckheere-Terpstra test to the multivariate case, which is based on

$$J = \frac{n^{-3/2}\sum_{h=1}^{k}J_h}{[\mathbf{1}_k \boldsymbol{\Upsilon} \mathbf{1}_k^T]^{1/2}}, \tag{4.3}$$

where J_h is the Jonckheere-Terpstra (Jonckheere, 1954; Terpstra, 1952) statistic computed on the hth variable and the covariance matrix $\boldsymbol{\Upsilon}$ has diagonal and off-diagonal elements $n^{-3}\text{Var}(J_h)$ and $n^{-3}\text{Cov}(J_h, J_g)$, respectively, for $h \neq g = 1, \ldots, k$. The asymptotic null distribution of J is standard normal.

Sim and Johnson (2004) used the maxmin criterion of Abelson and Tuckey (1963) under the assumption that $\boldsymbol{Y}_i \sim N_k(\boldsymbol{\mu}_i, \boldsymbol{\Sigma})$ with known $\boldsymbol{\Sigma}$. The test rejects the null hypothesis for large values of

$$L = \sum_{h=1}^{k}\sum_{g=1}^{k}\sigma^{hg}\left(\sum_{i=1}^{r}b_i\sum_{s=1}^{n_i}Y_{his}\right), \tag{4.4}$$

where the $\{\sigma^{hg}\}$, $h, g = 1, \ldots, k$, are the elements of $\boldsymbol{\Sigma}^{-1}$ and the b_i's, which satisfy $\sum_{i=1}^{r}n_ib_i = 0$, are the contrast coefficients given by

$$b_i = \frac{1}{n^{1/2}n_i}\left\{\sqrt{m_{i-1}(n - m_{i-1})} - \sqrt{m_i(n - m_i)}\right\}, \tag{4.5}$$

where $m_0 \equiv 0$ and $m_l = \sum_{i=1}^{l}n_i$, $l = 1, \ldots, r$. The null distribution of $L/[(\boldsymbol{\Sigma}^{-1}\mathbf{1}^T)^T\mathbf{1}^T\sum_{i=1}^{r}n_ib_i^2]^{1/2}$ is standard normal.

Under the assumption that $\boldsymbol{Y}_i \sim N_k(\boldsymbol{\mu}_i, \boldsymbol{\Sigma})$, one may rewrite the hypotheses by the reparametrization $\vartheta_{hl} = \mu_{hl} - \mu_{h(l-1)}$ for $l = 2, \ldots, r$. Thus the problem is rearranged in a one-sample setting as $H_0 : \bigcap_{h=1}^{k}\bigcap_{l=2}^{r}\{\vartheta_{hl} = 0\}$ against $H_1 : \bigcup_{h=1}^{k}\bigcup_{l=2}^{r}\{\vartheta_{hl} > 0\}$. Note that the dimensionality of the problem is now $q = k(r-1)$. The LRT based on a single observation $((\bar{\boldsymbol{Y}}_2 - \bar{\boldsymbol{Y}}_1) \ldots (\bar{\boldsymbol{Y}}_r - \bar{\boldsymbol{Y}}_{r-1}))$ from $N_q(\boldsymbol{\vartheta}, \boldsymbol{\Lambda})$, when $\boldsymbol{\Lambda}$ is assumed to be known, has null distribution chi-bar-square,

$$\Pr\{\bar{\chi}^2 \leq c | H_0\} = \sum_{h=0}^{q} w_h(q, \Lambda, I\!\!R^{+q}) \Pr\{\chi_h^2 \leq c\},$$

where χ_h^2 are central chi-square variables with h degrees of freedom ($h = 0, \ldots, q$ and $\chi_0^2 \equiv 0$) and weights $w_h(q, \Lambda, I\!\!R^{+q})$ that can be explicitly determined up to $q = 4$.

In order to keep the dimension more manageable in the r-sample multivariate setting, Sim and Johnson (2004) propose a likelihood ratio test using the contrast coefficients, that is based on

$$\text{LRT} = \frac{1}{\sum_{i=1}^{r} n_i b_i^2} \left\{ \mathbf{Z} \mathbf{\Sigma}^{-1} \mathbf{Z}^T - \min_{\theta \geq 0_k} \left[(\mathbf{Z} - \theta) \mathbf{\Sigma}^{-1} (\mathbf{Z} - \theta)^T \right] \right\}, \quad (4.6)$$

where the b_i's are given in (4.5), $\mathbf{Z} = \sum_{i=1}^{r} b_i \sum_{s=1}^{n_i} \mathbf{Y}_{is}$, and $\theta = E(\mathbf{Z}) = \sum_{i=1}^{r} b_i n_i \boldsymbol{\mu}_i$. When $\mathbf{\Sigma}$ is assumed to be known, LRT in (4.6) has null distribution $\Pr\{\text{LRT} \leq c\} = \sum_{h=0}^{k} w_h(k, \mathbf{\Sigma}, I\!\!R^{+k}) \Pr\{\chi_h^2 \leq c\}$.

Table 4.2. Serum enzyme levels for forty rats. Serum enzyme levels are in international units/liter. Dosage of vinylidene fluoride in parts/million.

		Rat within Dosage									
Dosage	Enzyme	1	2	3	4	5	6	7	8	9	10
0	SDH	18	27	16	21	26	22	17	27	26	27
	SGOT	101	103	90	98	101	92	123	105	92	88
1500	SDH	25	21	24	19	21	22	20	25	24	27
	SGOT	113	99	102	144	109	135	100	95	89	98
5000	SDH	22	21	22	30	25	21	29	22	24	21
	SGOT	88	95	104	92	103	96	100	122	102	107
15,000	SDH	31	26	28	24	33	23	27	24	28	29
	SGOT	104	123	105	98	167	111	130	93	99	99

The data in Table 4.2 are taken from Dietz (1998). In a dose-response experiment on vinylidene fluoride, a chemical suspected of causing liver damage, four groups of ten "Fischer-344" rats received by inhalation exposure increasing dosages. Among the response variables measured on the rats were two serum enzymes: SDH and SGOT. Increasing levels of these enzymes are often associated with liver damage. It is of interest to test whether each enzyme level stochastically increases with increasing doses of vinylidene fluoride. The data are shown in Table 4.2. By using $B = 5000$ permutations, we compute $T_{hi} = \sum_{i=i+1}^{r} \sum_{s=1}^{n_i} Y_{his}$ first, and then we obtain the Fisher combined test statistics $T_h = -2 \log(\prod_{i=1}^{r-1} p_{hi})$ and $T = -2 \log(\prod_{h=1}^{k} p_h)$, giving a p-value of 0.0004 for the global null hypothesis H_0, whereas the J statistic equals 2.72, significant at $\Pr\{N(0, 1) > 2.72\} = 0.0033$, the L statistic equals 2.886, significant at $\Pr\{N(0, 1) > 2.886\} = 0.0020$, and the LRT statistic equals 9.4225, significant at $\Pr\{\bar{\chi}^2 > 9.4225\} = 0.0041$.

4.4.1 Applications with R functions

For performing the statistical tests illustrated in this section, load the data from Table 4.2, where X represents class labels and Y the vector of data, and obtain the $B \times k \times (r-1)$ matrix of test statistics $T_{hi}^*(b)$, $b = 1, \ldots, B$, $h = 1, \ldots, k$, $i = 1, \ldots, r-1$:

```
> load("rats.Rdata")
> source("pstocRs.R")
> set.seed(0)
> B <- 5000
> T_hi <- pstocRs(X,Y,B)
```

Obtain the corresponding matrix of p-values

```
> load("t2p.R")
> P_hi <- T_hi
> r = length(unique(X))
> for (i in 1:(r-1)){ P_hi[,,i] <- t2p(T_hi[,,i]) }
```

and compute the Fisher combined test statistic $T_h = -2 \log(\prod_{i=1}^{r-1} p_{hi})$, obtaining the raw p-values for variables SDH and SGOT, respectively,

```
> T_h <- P_hi[,,1]
> T_h <- apply(P_hi,c(1,2),function(x){-2*log(prod(x))})
> P_h<-t2p(T_h)
> P_h[1,]
[1] 0.002 0.075
```

and finally compute the global test $T = -2 \log(\prod_{h=1}^{k} p_h)$ for testing H_0:

```
> T <- P_hi[,1,1]
> T <- apply(P_h,1,function(x){-2*log(prod(x))})
> sum(T[-1]>=T[1])/(B-1)
[1] 0.00040008
```

## Part II

Nonparametric ANOVA

Nonparametric One-Way ANOVA

The one-way ANOVA is a well-known testing problem for the equality in distribution of $C \geq 2$ groups of sampling data, where C represents the number of treatment levels in an experiment. In this framework, units belonging to the jth group, $j = 1, ..., C$, are presumed to receive treatment at the jth level. When side assumptions specific to the problem ensure that responses are homoscedastic, the equality of C distributions may be reduced to that of C mean values. Here we consider fixed effects in additive response models,

$$Y_{ij} = \mu + \delta_j + \sigma \varepsilon_{ij} \quad i = 1, \ldots, n_j; \quad j = 1, \ldots, C, \quad (5.1)$$

where μ is a population constant, δ_j are the fixed treatment effects that satisfy the contrast condition $\sum_j \delta_j = 0$, ε_{ji} are exchangeable random errors with zero mean value and unit scale parameter, σ is a scale coefficient that is assumed to be invariant with respect to groups, and C is the number of groups into which the data are partitioned. Note that, in this model, responses are assumed to be homoscedastic.

Observe that these conditions are equivalent to equality of C distributions: $H_0 : \{Y_1 \overset{d}{=} Y_2 \overset{d}{=} \ldots \overset{d}{=} Y_C\}$. Also note that this equality implies that data $\mathbf{Y}$ are exchangeable; in particular, they may be viewed as if they were randomly attributed to the groups.

In the case of normality, this problem is solved by Snedecor's well-known F test statistic:

$$F = \frac{(N - C) \sum_{j=1}^{C} n_j [\bar{Y}_j - \bar{Y}]^2}{(C - 1) \sum_{j=1}^{C} \sum_{i=1}^{n_j} [Y_{ij} - \bar{Y}_j]^2}. \quad (5.2)$$

Let us assume that we can maintain homoscedasticity and the null hypothesis in the form $H_0 : \{Y_1 \overset{d}{=} Y_2 \overset{d}{=} \ldots \overset{d}{=} Y_C\}$ but that we cannot maintain normality. Assuming the existence, in H_0, of a common nondegenerate, continuous, unknown distribution P, the problem may be solved by a rank test such as Kruskal-Wallis or by conditioning with respect to a set of sufficient statistics (i.e., by a permutation procedure). Note that, because of conditioning, the latter procedure allows for relaxation of continuity for P. The

D. Basso et al., *Permutation Tests for Stochastic Ordering and ANOVA*, Lecture Notes in Statistics, 194, DOI 10.1007/978-0-387-85956-9_5,
© Springer Science+Business Media, LLC 2009

permutation solution also allows for relaxation of some forms of homoscedasticity for responses in H_1. Indeed, the generalized one-way ANOVA model allowing for unbiased permutation solutions assumes that the hypotheses are

$$H_0 : \{Y_1 \stackrel{d}{=} Y_2 \stackrel{d}{=} \ldots \stackrel{d}{=} Y_C\} \text{ against } H_1 : \{Y_1 \stackrel{d}{\neq} Y_2 \stackrel{d}{\neq} \ldots \stackrel{d}{\neq} Y_C\},$$ with the restriction that, for every pair $h \neq j$, $h, j = 1, \ldots, C$, the corresponding response variables are stochastically ordered (pairwise dominance relationship) according to either $Y_h \stackrel{d}{>} Y_j$ or $Y_h \stackrel{d}{<} Y_j$. Thus, $\forall\ y \in \mathbb{R}$, and the associated c.d.f.s are related according to either $F_h(y) \leq F_j(y)$ or $F_h(y) \geq F_j(y)$.

This pairwise dominance assumption may correspond to a model in which treatment may affect both location and scale coefficients, as for instance in $\{Y_{ji} = \mu + \delta_j + \sigma(\delta_j) \cdot \varepsilon_{ij}, i = 1, \ldots, n_j, j = 1, \ldots, C\}$, where $\sigma(\delta_j)$ are monotonic functions of treatment effects δ_j or their absolute values $|\delta_j|$, provided that $\sigma(0) = \sigma$ and pairwise stochastic ordering on c.d.f.s is preserved. The latter model is consistent with the notion of randomization. Therefore, (i) in the randomization context, units are assumed to be randomly assigned to treatment levels, so that H_0 implies exchangeability of responses; (ii) in the alternative, treatment may jointly affect location and scale coefficients, so that the resulting permutation distributions become either stochastically larger or smaller than the null ones. Also note that the pairwise dominance assumption is consistent with a generalized model with stochastic effects of the form $\{Y_{ji} = \mu + \sigma \cdot \varepsilon_{ij} + \Delta_{ji}, i = 1, \ldots, n_j, j = 1, \ldots, C\}$, where Δ_{ji} are the stochastic treatment effects that satisfy the (pairwise) ordering condition that for every pair $h \neq j$, $i, j = 1, \ldots, C$, we have either $\Delta_h \stackrel{d}{>} \Delta_j$ or $\Delta_h \stackrel{d}{<} \Delta_j$.

5.1 Overview of Nonparametric One-Way ANOVA

Formalizing the testing problem for a C-sample one-way ANOVA, we assume that $\mathbf{Y} = \{\mathbf{Y}_1, \ldots, \mathbf{Y}_C\}$ represents the data set partitioned into C groups, where $\mathbf{Y}_j = \{Y_{ji}, i = 1, \ldots, n_j\}$, $j = 1, \ldots, C$, are i.i.d. observations from nondegenerate distributions P_j. We also assume that the sampling means are proper indicators of treatment effects. Under the response (homoscedastic) model (5.1), the hypotheses are

$$H_0 : \{Y_1 \stackrel{d}{=} \ldots \stackrel{d}{=} Y_C\} = \{\delta_1 = \ldots = \delta_C = 0\}$$

against $H_1 : \{$at least one equality is not true$\}$. If it is suitable for analysis, we may consider a data transformation φ, so that related sampling means become proper indicators for treatment effects. According to the CMC procedure, iterations are now done from the pooled data set $\mathbf{Y} = [\mathbf{Y}_1, \ldots, \mathbf{Y}_C]$, which is a set of sufficient statistics for the problem in H_0. If associative test statistics are used, the related permutation sample space $\mathcal{Y}_{/\mathbf{Y}}$ contains $N!/(n_1! \cdot \ldots \cdot n_C!)$ distinct points, where $N = \sum_j n_j$ is the total sample size.

According to the assumptions above, a suitable test is the Kruskal-Wallis rank test, which is based on the statistic

$$\mathrm{KW} = \left\{ 12 \cdot \sum_{j=1}^{C} n_j \cdot \left[\bar{R}_j - (n+1)/2 \right]^2 \right\} \cdot \frac{1}{N(N+1)},$$

where R_{ji} is the rank of Y_{ji}, $j = 1, \ldots, C$, $i = 1, \ldots, n_j$, within the pooled data set $\mathbf{Y}$, and $\bar{R}_j = \sum_i R_{ji}/n_j$, $j = 1, \ldots, C$, is the jth mean rank. For large sample sizes n_j, the null distribution of KW is approximated by that of a central χ^2 with $C - 1$ degrees of freedom.

The problem of testing the null hypothesis $H_0 : \{P_1 = \ldots = P_C\}$ (the C groups have the same distribution) against the alternative $H_1 : \{H_0$ is not true$\}$ (at least one distribution is different from the others) may be solved, for example, by an Anderson-Darling type test, which is based on the statistic

$$T_{AD}^{*2} = \sum_{j=1}^{C} \sum_{i=1}^{n_j} \left[F_j^*(Y_{(i)j}^*) - \hat{F}.(Y_{(i)j}^*) \right]^2 \cdot \left(\hat{F}.(Y_{(i)j}^*)[1 - \hat{F}.(Y_{(i)j}^*)] \right)^{-1},$$

where $Y_{(i)j}^*$ are the permutation order statistics in the jth group, $\hat{F}.(y) = \sum_j n_j \cdot \hat{F}_j(y)/n$ is the pooled empirical distribution function, and all the other symbols have obvious meanings.

5.2 Permutation Solution

In order to solve the one-way ANOVA problem, let us presume that side assumptions are such that the response data behave according to the additive model (5.1) and that treatment effects satisfy the constraint $\sum_j \delta_j = 0$. In particular, when $\sigma(\delta_j) = \sigma$, we have the so-called *homoscedastic* situation. Of course, without loss of generality, the hypotheses for this problem become $H_0 : \{\delta_1 = \ldots = \delta_C = 0\}$ against $H_1 : \{H_0$ is not true$\}$, and a suitable test statistic is given by (5.2). Therefore, in order to obtain a permutation test, the following algorithm applies:

- Compute the test statistic (5.2) from the observed data $\mathbf{y} = [\mathbf{y_1}, \mathbf{y_2}, \ldots, \mathbf{y_C}]$,

$$F = \frac{(N - C) \sum_{j=1}^{C} n_j [\bar{y}_j - \bar{y}]^2}{(C - 1) \sum_{j=1}^{C} \sum_{i=1}^{n_j} [y_{ij} - \bar{y}_j]^2}.$$

- Select a large number of permutations B. Then, for $b = 1, \ldots, B$, repeat:
  1. Obtain a permutation of the whole data vector $\mathbf{y}^* = \pi_b \mathbf{y}$, where π_b is the bth random permutation.

2. Compute the test statistic (5.2) from the observed data,

$$F_b^* = \frac{(N-C)\sum_{j=1}^{C} n_j[\bar{y}_j^* - \bar{y}]^2}{(C-1)\sum_{j=1}^{C}\sum_{i=1}^{n_j}[y_{ij}^* - \bar{y}_j^*]^2}.$$

• Compute the p-value of the test as

$$p = \frac{\#[F_b^* \geq F]}{B}.$$

Note that, being a permutation test, the constant $(N-C)/(C-1)$ can be omitted. Moreover, we have, for any permutation,

$$F_b^* \sim \frac{\sum_{j=1}^{C} n_j[\bar{y}_j^* - \bar{y}]^2}{\sum_{j=1}^{C}\sum_{i=1}^{n_j}[y_{ij}^* - y] - \sum_{j=1}^{C} n_j[\bar{y}_j^* - \bar{y}]^2}$$

$$= \frac{T(\mathbf{y}^*)}{K - T(\mathbf{y}^*)},$$

where $K = \sum_j \sum_i (y_{ij}^* - \bar{y})^2$ is constant and the symbol "$\sim$" means *proportional to*. Therefore, the test statistic (5.2) is permutationally equivalent to (i.e., it is a nondecreasing monotonic function of) the test statistic

$$T(\mathbf{y}) = \sum_{j=1}^{C} n_j[\bar{y}_j - \bar{y}]^2. \tag{5.3}$$

This means that, for every sample $\mathbf{y} \in \mathcal{Y}$, we can apply the algorithm of this section with (5.2) replaced by (5.3) and obtain to the same inferential conclusions. If the null hypothesis is rejected, it is often of interest to investigate in detail which experimental group(s) led to its rejection. This is usually done by the parametric *post-hoc comparison*, like *Tukey's honest significant difference* (HSD).

In post-hoc comparisons, a pairwise testing procedure is applied to all possible pairs of groups (i.e., there are $C(C-1)/2$ hypotheses of interest: $H_0^{jk} : \delta_j = \delta_k, 1 \leq j < k \leq C$). Thus, a collection of two-sample tests for any specific hypothesis is performed, and a multiplicity problem arises because all tests are done simultaneously. The nonparametric combination introduced in Chapter 1 allows us to view the test for one-way ANOVA problem as a combination of partial tests for pairwise comparisons. To see this, note that (5.3) is permutationally equivalent to

$$T'(\mathbf{y}) = \sum_{j=1}^{C}\sum_{k\neq j}^{C} n_j n_k[\bar{y}_j - \bar{y}_k]^2, \tag{5.4}$$

To see this, recall the deviance decomposition

$$\sum_{j=1}^{C}\sum_{k\neq j}^{C} n_j n_k [\bar{y}_j - \bar{y}_k]^2 = 2\sum_{j=1}^{C}\sum_{k\neq j}^{C} n_j n_k \bar{y}_j^2 - 2\sum_{j=1}^{C} n_j \bar{y}_j \sum_{k\neq j}^{C} n_k \bar{y}_k$$

$$= 2\sum_{j=1}^{C}(N - n_j)n_j \bar{y}_j^2 - 2\sum_{j=1}^{C} n_j \bar{y}_j [N\bar{y} - n_j \bar{y}_j]$$

$$= 2\sum_{j=1}^{C}(N - n_j)n_j \bar{y}_j^2 - 2N^2 \bar{y}^2 + 2\sum_{j=1}^{C} n_j^2 \bar{y}_j^2$$

$$= 2N\sum_{j=1}^{C} n_j [\bar{y}_j^2 - \bar{y}]^2,$$

where $N = \sum_{k=1}^{C} n_k$. The test statistic (5.4) is a *direct* combination of $C(C - 1)/2$ partial test statistics T_{jk}^2, where

$$T_{jk} = \sqrt{n_j n_k}[\bar{y}_j - \bar{y}_k]. \tag{5.5}$$

Note that T_{jk}^2 is suitable for assessing H_0^{jk} against the alternative $H_1^{jk} : \delta_j \neq \delta_k$. Now the global null hypothesis can be viewed as an intersection of partial null hypotheses,

$$H_0 = \{\delta_1 = \ldots = \delta_C = 0\} = \bigcap_{j\neq k} H_0^{jk}.$$

Under H_0, the response elements are exchangeable, and this allows us to simultaneously perform the partial tests to assess $H_0^{jk} : \delta_j = \delta_k$, $1 \leq j < k \leq C$, by computing the statistics (5.5) at each permutation of the response. Large values of each partial test are significant against H_0^{jk}, and large values of their direct combination are significant against H_0. We can therefore simultaneously obtain partial tests for post-hoc comparisons and a global test for the one-way ANOVA problem. The null distribution of partial tests is obtained by computing, at each permutation, the value of the test statistics (5.5)

$$T_{jk}^{*2} = n_j n_k [\bar{y}_j^* - \bar{y}_k^*]^2 \qquad 1 \leq j < k \leq C, \tag{5.6}$$

and the partial p-values by computing

$$p_{jk} = \frac{\#[T_{jk}^{*2} \geq T_{jk}^2]}{B} \qquad 1 \leq j < k \leq C.$$

The null distribution of the global test is obtained by computing, at each permutation, the statistic

$$T'^* = \sum_{j=1}^{C}\sum_{k\neq j}^{C} T_{jk}^{*2} = \sum_{j=1}^{C}\sum_{k\neq j}^{C} n_j n_k [\bar{y}_j^* - \bar{y}_k^*]^2,$$

and the global p-value is obtained accordingly. If the global p-value is significant, then one can look at the partial p-values in order to do the post-hoc comparisons. Since the partial p-values are done simultaneously, a correction for multiplicity is required.

5.2.1 Synchronizing Permutations

Post-hoc comparisons represent the further step of any accurate statistical analysis if the H_0 is rejected. But, if one wants to perform post-hoc comparisons through permutation tests, some care is needed. Actually, there are several ways to permute data. In this section we investigate some proposals and compare them in order to suggest the "best" way of permuting the response, at least in balanced designs.

The fisrt proposal is suggested by the global null hypothesis: Since under H_0 the observations of all groups are exchangeable, we can apply *pooled* permutations involving the whole data vector. However, if we are interested in multiple comparisons, then a further aspect is to be considered: Each comparison should only involve the observations belonging to the pair of groups considered; for example, while comparing the jth and kth groups, the inference should be made on the *paired* vector $\mathbf{y_{jk}} = [\mathbf{y_j}', \mathbf{y_k}']'$ independently. This is because if we permute the whole vector of data, observations from possibly active groups (i.e., with $\delta_l \neq 0$, $l \neq j, k$) could influence the partial p-value p_{jk}.

Another aspect to take into account is that the partial tests T_{jk}'s are dependent. Indeed, let $C = 3$, and consider the balanced case with $n_j = n \ \forall \ j$: For any permutation, we have the equivalence

$$T_{12}^* = T_{13}^* - T_{23}^*.$$

If the design is balanced, a further possibility is to apply *synchronized* permutations (see Chapter 6). They can be obtained by permuting the rows of the pseudodata matrix

$$\mathbf{Y} = \begin{bmatrix} \mathbf{y_1} & \mathbf{y_1} & \mathbf{y_2} \\ \hline \mathbf{y_2} & \mathbf{y_3} & \mathbf{y_3} \end{bmatrix} = \begin{bmatrix} y_{11} \\ y_{21} \\ \vdots \\ y_{n1} \\ y_{12} \\ y_{22} \\ \vdots \\ y_{n2} \end{bmatrix} \begin{bmatrix} y_{11} \\ y_{21} \\ \vdots \\ y_{n1} \\ y_{13} \\ y_{23} \\ \vdots \\ y_{n3} \end{bmatrix} \begin{bmatrix} y_{12} \\ y_{22} \\ \vdots \\ y_{n2} \\ y_{13} \\ y_{23} \\ \vdots \\ y_{n3} \end{bmatrix}.$$

We have already shown in Chapter 1 that the NPC methodology accounts for dependencies among partial statistics. This is done by obtaining the joint (permutation) null distribution of partial statistics. As a consequence, paired permutations should be done not independently but jointly. Applying synchronized permutations allows both maintaining the dependencies among partial tests and involving the observations of each comparison at the same time. Let $\mathbf{Y}^*$ be a row permutation of $\mathbf{Y}$. Then the joint distribution of partial statistics is given by

$$[T_{12}^*, T_{13}^*, T_{23}^*] = [\mathbf{1_n}', -\mathbf{1_n}']\mathbf{Y}^*,$$

where $\mathbf{1_n}$ is an $n \times 1$ vector of 1s.

Figures 5.1, 5.2 and 5.3 show the differences in the joint distribution of partial statistics when applying pooled, paired, and synchronized permutationspermutation!synchronized, respectively. Note how the inner dependencies among partial tests are maintained in Figures 5.1 and 5.3, but not in Figure 5.2.

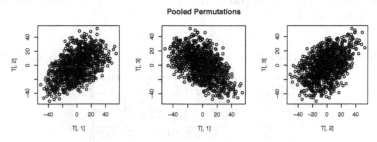

Fig. 5.1. Joint distribution of $[T_{12}^*, T_{13}^*, T_{23}^*]$ when permuting the whole data vector.

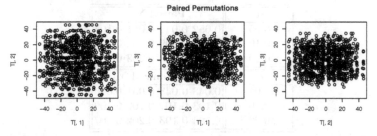

Fig. 5.2. Joint distribution of $[T_{12}^*, T_{13}^*, T_{23}^*]$ when permuting $\mathbf{y_{jk}}$ independently.

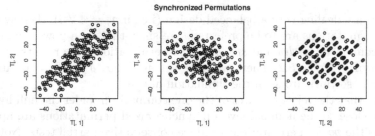

Fig. 5.3. Joint distribution of $[T_{12}^*, T_{13}^*, T_{23}^*]$ with synchronized permutations.

Table 5.1 reports the rejection rates of partial and global tests that are related to Figures 5.1, 5.2, and 5.3. The simulation is carried on under H_0, with $C = 3$, $n_j = 5$, and normally distributed errors. Note that partial tests for multiple comparisons achieve the nominal level α whatever permutations are applied, but the global test T' is anticonservative when *paired* permutations are applied. This is because the dependence among partial tests is not taken into account here, and therefore there are some permutation points that do not belong to the domain of the global test statistic T'. These simulations also show that the global test allows controlling of the familywise error (FWE) in the weak sense. *Pooled* permutations are suitable in the unbalanced case, although the related global test is conservative. *Synchronized* permutations, on the contrary, allow exact testing for both partial and global null hypotheses. The limitation about applying synchronized permutations is the cardinality of the support of the global test statistic when $n_j \equiv n$ is small. See Section 6.3 for further details.

Table 5.1. Rejection rates of H_0^{jk} and H_0 under H_0; $C = 3$, $n_j = 5$, $\varepsilon_{ij} \sim N(0,1)$.

Comparison	Δ_{1-2}	Δ_{1-3}	Δ_{2-3}	T'
T.E.	0	0	0	-
α	Pooled $\mathbf{Y}^*$			
0.05	0.043	0.049	0.047	0.040
0.10	0.092	0.096	0.089	0.088
0.20	0.180	0.187	0.190	0.185
α	Paired $\mathbf{Y}^*$			
0.05	0.052	0.049	0.051	0.105
0.10	0.097	0.099	0.104	0.152
0.20	0.208	0.208	0.203	0.239
α	Synchronized $\mathbf{Y}^*$			
0.05	0.048	0.048	0.051	0.046
0.10	0.103	0.102	0.100	0.109
0.20	0.194	0.193	0.202	0.207

If we are dealing with a balanced design (i.e., if $n_j = n \; \forall \; j$), then synchronized permutations are to be preferred. We show that by a power simulation study comparing *pooled* and *synchronized* permutations, the results of which are shown in Table 5.2. There we have set $\delta_1 = 2\sigma$, $\delta_2 = \delta_3 = 0$, and the remaining settings are as in the simulation above.

Note that the rejection rates of the one comparison under the null hypothesis are close to the nominal levels if synchronized permutations are applied, whereas the *pooled* permutations produce conservative partial tests. Note also that the rejection rates of the global null hypothesis are far bigger than the nominal levels.

Table 5.2. Rejection rates of partial and global null hypotheses under H_0.

	pooled $\mathbf{Y}^*$				Synchronized $\mathbf{Y}^*$			
True Δ	Δ_{1-2}	Δ_{1-3}	Δ_{2-3}	T'	Δ_{1-2}	Δ_{1-3}	Δ_{2-3}	T'
α	2σ	2σ	0	-	2σ	2σ	0	-
0.05	0.714	0.776	**0.012**	0.838	0.736	0.752	**0.054**	0.810
0.10	0.856	0.872	**0.028**	0.920	0.872	0.864	**0.096**	0.908
0.20	0.946	0.950	**0.080**	0.966	0.942	0.934	**0.202**	0.960

For these reasons, we suggest applying synchronized permutations at least when the design is balanced and n is not too small (say $n \geq 4$). Otherwise there is no guarantee that the nominal significance levels will be achieved.

5.2.2 A Comparative Simulation Study for One-Way ANOVA

The permutation test for one-way ANOVA and post-hoc comparisons has been evaluated and compared with other tests through a simulation study. The global permutation test (5.4) is compared with the nonparametric Kruskal-Wallis test and the parametric F test. The partial permutation tests have been compared with Tukey's HSD. The simulations refer to an experimental design with $C = 5$ groups and three observations per group. Data are normally distributed with location-shift vector $\boldsymbol{\delta} = [\delta_1, \ldots, \delta_5]$ and variance $\sigma^2 = 1$. All simulations were run with 1000 MC data generations. In the simulations of this section, we have applied the *pooled* permutations since here the cardinality of the support of the global test statistic is high enough to guarantee that the usual nominal significance levels are achievable ($\#[\mathcal{T}] = [nC]!$).

The first scenario refers to the null hypothesis (with $\boldsymbol{\delta} = [0,0,0,0,0]'$), whereas the second refers to a location vector $\boldsymbol{\delta} = [0,1,0,1,0]$. Table 5.3 reports the results of the one-way ANOVA test for the permutation test and the competitors considered. Under H_0 (top of the table), the global test performs very similarly to the F test, whereas the Kruskal-Wallis test seems to be more conservative. This fact may be due to the small sample size (the null distribution of the Kruskall-Wallis test is asymptotic). As regards the power comparisons, the permutation test is again very close to the F test performance.

Table 5.4 shows in detail the post-hoc comparison tests. The left side of the table refers to the simulation under H_0, and the right side concerns the power simulation with $\boldsymbol{\delta} = [0,1,0,1,0]'$. Since the R function we provide computes nonadjusted p-values (see Subsection 5.7.1), the nominal significance levels of pairwise comparisons are corrected by applying the Bonferroni adjustment. As regards Tukey's HSD, instead we applied the R function Tukey.HSD, which provides adjusted p-values, which can be directly compared with the nominal α levels.

Table 5.3. Rejection rates: global permutation, Kruskal-Wallis, and F tests. $C = 5$, $n_j \equiv 3$, and standard normal errors.

	$\delta = [0,0,0,0,0]'$		
α	T'	KW	F
0.05	0.056	0.024	0.057
0.10	0.108	0.090	0.110
0.20	0.205	0.212	0.211

	$\delta = [0,1,0,1,0]'$		
α	T'	KW	F
0.05	0.184	0.091	0.184
0.10	0.300	0.226	0.294
0.20	0.433	0.447	0.434

Table 5.4. Rejection rates: partial permutation tests and Tukey's HSD. $C = 5$, $n_j \equiv 3$. Standard normal errors.

	$\delta = [0,0,0,0,0]'$						$\delta = [0,1,0,1,0]'$					
	Partial tests			Tukey's HSD			Partial tests			Tukey'sHSD		
α	.005	.01	.02	.05	.1	.2	.005	.01	.02	.05	.1	.2
1-2	.007	.014	.022	.011	.025	.039	**.025**	**.044**	**.073**	**.047**	**.087**	**.142**
1-3	.005	.009	.022	.011	.018	.045	.005	.011	.017	.013	.024	.046
1-4	.003	.010	.019	.010	.019	.035	**.030**	**.048**	**.077**	**.054**	**.097**	**.152**
1-5	.007	.015	.023	.009	.018	.046	.002	.009	.019	.010	.026	.038
2-3	.004	.008	.017	.008	.016	.032	**.026**	**.040**	**.072**	**.055**	**.090**	**.136**
2-4	.005	.007	.018	.011	.018	.045	.000	.001	.008	.010	.022	.043
2-5	.008	.016	.022	.012	.024	.056	**.031**	**.037**	**.070**	**.050**	**.084**	**.140**
3-4	.002	.012	.023	.010	.023	.045	**.031**	**.052**	**.083**	**.058**	**.088**	**.152**
3-5	.005	.008	.018	.009	.017	.043	.003	.005	.013	.012	.020	.043
4-5	.006	.012	.021	.012	.020	.044	**.025**	**.036**	**.069**	**.051**	**.085**	**.150**

The group pairs are reported in the first column of Table 5.4. Note how under H_0 the rejection rates of the permutation partial tests are close to the nominal levels, whereas Tukey's HSD is conservative. In the power simulation columns, the comparisons under the alternative hypothesis H_1^{jk} are highlighted in bold-face. Note how the permutation partial tests can detect the δ-shifts in the alternative H_1^{jk}, whereas Tukey's HSD recognizes none.

5.3 Testing for Umbrella Alternatives

In many experimental situations, such as evaluating marginal gain in performance efficiency as a function of degree of training, drug effectiveness as a function of time, or crop yield as a function of the quantity of fertilizer

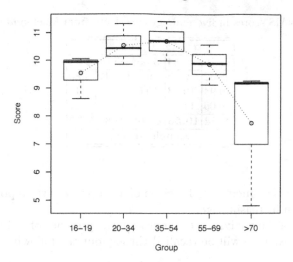

Fig. 5.4. Boxplot representation of the example from Mack and Wolfe (1981).

applied (to name but a few), the main interest is on evaluating the null hypothesis for one-way ANOVA against a specific alternative, which is usually known as the umbrella alternative. With the same symbols as in previous sections, let $\mathbf{Y} = [\mathbf{Y}_1, \mathbf{Y}_2, \ldots, \mathbf{Y}_C]'$ be C independent random samples, with $\mathbf{Y}_j = [Y_{1j}, Y_{2j}, \ldots, Y_{n_j j}]$, $j = 1, \ldots, C$ having continuous distribution function $F_j(y)$. For such C sample data, we are often interested in testing the null hypothesis that all C samples came from a single common distribution,

$$H_0 : F_1(y) = F_2(y) = \cdots = F_C(y), \qquad \forall y \in \mathbb{R}, \tag{5.7}$$

against the umbrella alternative hypothesis,

$$H_1 : F_1(y) \geq \cdots \geq F_{k-1}(y) \geq F_k(y) \leq F_{k+1}(y) \leq \cdots \leq F_C(y), \tag{5.8}$$

with at least one strict inequality for at least one y value. We refer to these as *umbrella alternatives* because of the configuration of the corresponding population means and call k the point peak of the umbrella (see Figure 5.4). Note that, if the peak group is known to be equal to $\hat{k}$, then H_1 may be decomposed as:

$$H_1 : \{F_1(y) \geq \cdots \geq F_{\hat{k}-1}(y) \geq F_{\hat{k}}(y)\} \bigcap \{F_{\hat{k}}(y) \leq F_{\hat{k}+1}(y) \leq \cdots \leq F_C(y)\};$$

that is, the alternative hypothesis may be viewed as the intersection of two simple stochastic ordering alternatives.

Let us introduce the umbrella alternative problem by an example that appeared in Mack and Wolfe (1981, p. 178). The data of Table 5.5 are values in the range typically obtained on the Welchsler Adult Intelligence Scale (WAIS) by males of various ages. It is generally believed that the ability to comprehend

Table 5.5. WAIS scores in five male groups. Data from Mack and Wolfe (1981).

Age Group				
15-19	20-34	35-54	55-69	>70
8.62	9.85	9.98	9.12	4.80
9.94	10.43	10.69	9.89	9.18
10.06	11.31	11.4	10.57	9.27
9.54	10.53	10.69	9.86	7.75
Sample Means				

ideas and learn is an increasing function of age up to a certain point, and then it declines with increasing age.

The scope of the analysis is to determine the presence of a significant peak (if any). This example will be recalled throughout the following sections.

5.4 Simple Stochastic Ordering Alternatives

Under the assumption of model (5.1), let us consider the simple stochastic ordering problem for the first $\hat{k}$ samples to assess the null hypothesis $F_1(y) = F_2(y) = \cdots = F_{\hat{k}}(y) \; \forall \; y \in I\!\!R$ against the alternative $F_1(y) \geq \cdots \geq F_{\hat{k}-1}(y) \geq F_{\hat{k}}(y)$. Note that, under the null hypothesis, the elements of the response are exchangeable (this fact enables us to provide the null distribution of a proper test statistic).

If $\hat{k} = 2$, the stochastic ordering problem reduces to a two-sample problem with a restricted alternative. If $\hat{k} > 2$, then let us consider the whole data set to be split into two pooled pseudo-groups, where the first is obtained by pooling together data of the first j (ordered) groups and the second by pooling the rest. In order to better understand the reason why we pool together the ordered groups, suppose $\hat{k} = 3$ and let us consider the following theorem.

Theorem 5.1. *Let X_1, X_2, X_3 be mutually independent random variables that admit cumulative distribution function $F_j(t)$, $t \in I\!\!R$, $j = 1, 2, 3$. Then, if $X_1 \overset{d}{\leq} X_2 \overset{d}{\leq} X_3$, we have*

$$(i) \; X_1 \overset{d}{\leq} X_2 \oplus X_3 \qquad and \qquad (ii) \; X_1 \oplus X_2 \overset{d}{\leq} X_3,$$

where $W \oplus V$ indicates a mixture of random variables W and V.

Proof. By definition, $X_1 \overset{d}{\leq} X_2 \overset{d}{\leq} X_3$ is equivalent to $F_1(t) \geq F_2(t) \geq F_3(t)$, $t \in I\!\!R$. The random variable $X_1 \oplus X_2$ has a cumulative distribution function equal to

$$F_{X_1 \oplus X_2}(t) = \omega_1 F_1(t) + \omega_2 F_2(t) \qquad \text{with } \omega_1 + \omega_2 = 1,$$

and $\omega_j > 0$, $j = 1, 2$. Therefore, by hypothesis,

$$F_{X_1 \oplus X_2}(t) = \omega_1 F_1(t) + \omega_2 F_2(t)$$
$$\geq \omega_1 F_2(t) + \omega_2 F_2(t) = F_2(t),$$

so $X_1 \oplus X_2 \overset{d}{\leq} X_2$ and we have proved (ii). In the same way, let $F_{X_2 \oplus X_3}(t) = \omega_2 F_2(t) + \omega_3 F_3(t)$ with $\omega_2 + \omega_3 = 1$ and $\omega_j > 0$, $j = 2, 3$. Then

$$F_{X_2 \oplus X_3}(t) = \omega_2 F_2(t) + \omega_3 F_3(t)$$
$$\leq \omega_2 F_2(t) + \omega_3 F_2(t) = F_2(t),$$

therefore $X_2 \oplus X_3 \overset{d}{\geq} X_2$, and this proves (i).

Now, conditional on the observed data, consider the pooled vector of observations $\mathbf{y_1} \uplus \mathbf{y_2} = [\mathbf{y_1}, \mathbf{y_2}]'$, where $\mathbf{y_j}$ is a vector of n_j observations from Y_j, $j = 1, 2$. Then $Y_1 \oplus Y_2$ has (empirical) cumulative distribution function equal to

$$\hat{F}_{Y_1 \oplus Y_2}(y) = \frac{1}{n_1 + n_2} \#[Y_{ij} \leq y]$$
$$= \frac{n_1}{n_1 + n_2} \frac{\#[Y_{i1} \leq y]}{n_1} + \frac{n_2}{n_1 + n_2} \frac{\#[Y_{\ell 2} \leq y]}{n_2}$$
$$= \omega_1 \hat{F}_{Y_1}(y) + \omega_2 \hat{F}_{Y_2}(y),$$

so the random variable describing the pooled vector of observations $\mathbf{Y_1} \uplus \mathbf{Y_2}$ has a mixture distribution. By extending this result to the $\hat{k}$ groups and by applying Theorem 5.1, we have that if $Y_1 \overset{d}{\leq} Y_2 \overset{d}{\leq} \cdots \overset{d}{\leq} Y_{\hat{k}}(y)$ holds, then

$$Y_{1 \oplus 2 \oplus \cdots \oplus j} \overset{d}{\leq} Y_{j+1 \oplus j+2 \oplus \cdots \oplus \hat{k}} \qquad \forall j \in \{1, \ldots, \hat{k} - 1\}.$$

In general, let $\mathbf{Z}_{(1)j} = [\mathbf{Y_1}, \mathbf{Y_2}, \ldots, \mathbf{Y_j}]'$ be the first (ordered) pseudo-group and let $\mathbf{Z}_{(2)j} = [\mathbf{Y_{j+1}}, \ldots, \mathbf{Y_{\hat{k}}}]'$ be the second (ordered) pseudo-group, $j = 1, \ldots, \hat{k} - 1$. In the null hypothesis, data of every pair of pseudo-groups are exchangeable because related pooled variables satisfy the relationships $Z_{1(j)} \overset{d}{=} Z_{2(j)}$, $j = 1, \ldots, \hat{k} - 1$. In the alternative, by Theorem 5.1, we have $Z_{1(j)} \overset{d}{\leq} Z_{2(j)}$, which corresponds to the monotonic stochastic ordering (dominance) between any pair of pseudo-groups (i.e., for $j = 1, \ldots, \hat{k} - 1$). This suggests that we express the hypotheses in the equivalent form

$$H_0 : \left\{ \bigcap_{j=1}^{\hat{k}-1} (Z_{1(j)} \overset{d}{=} Z_{2(j)}) \right\}$$

against

$$H_{1\hat{k}}^{\nearrow} : \left\{ \bigcup_{j=1}^{\hat{k}-1} (Z_{1(j)} \overset{d}{\leq} Z_{2(j)}) \right\},$$

where a breakdown into a set of sub hypotheses is emphasized.

Let us pay attention to the jth sub hypothesis $H_{0j} : \{Z_{1(j)} \overset{d}{=} Z_{2(j)}\}$ against $H_{1j} : \{Z_{1(j)} \overset{d}{\leq} Z_{2(j)}\}$. Note that the related sub-problem corresponds to a two-sample comparison for restricted alternatives, a problem that has an exact and unbiased permutation solution (for further details see Pesarin, 2001). This solution is based on the test statistics (among others)

$$T_{j\nearrow} = \bar{Z}_{2(j)} - \bar{Z}_{1(j)} \qquad j = 1, \ldots, \hat{k} - 1,$$

where $\bar{Z}_{2(j)}$ and $\bar{Z}_{1(j)}$ are sample means of the second pseudo-group and the first pseudo-group, respectively. The test statistics $T_{j\nearrow}$ are significant for large values. We can obtain a permutation test for each sub problem $H_{0(j)}$ vs. $H_{1(j)}^{\nearrow}$ by the following algorithm:

- Let $\mathbf{y} = [\mathbf{y}_1, \mathbf{y}_2, \ldots, \mathbf{y}_{\hat{k}}]'$ be the whole observed data vector of $\hat{k}$ groups.
- For $j = 1, \ldots, \hat{k} - 1$, repeat:
  1. Let $\mathbf{z}_{1(j)} = [\mathbf{y}_1, \ldots, \mathbf{y}_j]'$ and $\mathbf{z}_{2(j)} = [\mathbf{y}_{j+1}, \ldots, \mathbf{y}_{\hat{k}}]'$.
  2. Compute the observed values of the partial test statistics for the sub-problem $H_{0(j)}$ vs. $H_{1(j)}^{\nearrow}$ by computing

$$T_{j\nearrow} = \bar{z}_{2(j)} - \bar{z}_{1(j)}. \tag{5.9}$$

- Consider a large number B of random permutations of the response $\mathbf{y}$, and let $\mathbf{y}_b^*$ be a random permutation of $\mathbf{y}$. At each step $b = 1, \ldots, B$, repeat:
  1. Let $z_{1(j)}^*$ be the pseudo-vector with the first $n_{1(j)} = \sum_{\ell=1}^{j} n_\ell$ observations and $z_{2(j)}^*$ be the vector of the last $n_{2(j)} = \sum_{\ell=j+1}^{\hat{k}} n_\ell$ observations of $\mathbf{y}_b^*$.
  2. Obtain the permutation null distribution of the test statistic by computing

$$T_{j\nearrow}^* = \bar{z}_{2(j)}^* - \bar{z}_{1(j)}^*.$$

- Obtain the p-value of each sub - problem (partial p-value) by computing

$$p_{j\nearrow} = \frac{\#[T_{j\nearrow}^* \geq T_{j\nearrow}]}{B}$$

The previous algorithm provides $\hat{k} - 1$ p-values related to the sub hypotheses $H_{0(j)}$ against $H_{1(j)}^{\nearrow}$. In order to combine the partial information into a global test we require the NPC methodology, which was introduced in Section 1.4. Obvioulsy, if the alternative hypothesis is

$$H_{1\hat{k}}^{\searrow} : F_{\hat{k}}(y) \leq F_{\hat{k}+1}(y) \cdots \leq F_K(y),$$

the previous algorithm still applies by replacing the test statistic (5.9) with

$$T_{j\searrow} = Z_{1(j)} - Z_{2(j)}.$$

5.5 Permutation Test for Umbrella Alternatives

If the peak group is known, then the umbrella alternative can be detected by combining together two partial tests for simple stochastic ordering alternatives. Since finding the peak is the aim of the study, it will generally be unknown. However, we can detect the peak group by repeating the procedure for a known peak *as if* every group were the known peak group; that is, for each $k \in 1, \ldots, C$, let

$$\psi_{k\nearrow} = \sum_{j=1}^{k} T_{j\nearrow} \quad \text{and} \quad \psi_{k\searrow} = \sum_{j=k}^{C-1} T_{j\searrow}$$

be two partial tests to assess $H_{0k} : F_1(y) = F_2(y) = \cdots = F_C(y)$, $y \in \mathbb{R}$, against $H_{1k}^{\nearrow}$ and $H_{1k}^{\searrow}$, respectively, by applying the *direct* nonparametric combination of the partial tests $T_{j\nearrow}$'s and $T_{j\searrow}$'s. Then:

- Obtain the null distribution of the p-values to assess H_{0k} against $H_{1k}^{\nearrow}$ and $H_{1k}^{\searrow}$, respectively, say the pair $({}^{b}p_{k\nearrow}^{G*}, {}^{b}p_{k\searrow}^{G*})$, $b = 1, \ldots, B$. Let $(p_{k\nearrow}^{G}, p_{k\searrow}^{G})$ be the pair of p-values from the observed data (i.e., with $\mathbf{y}^* = \mathbf{y}$).
- Obtain the observed value of the test statistic with *Fisher's* NPC function:

$$\Psi_k = -2\log(p_{k\nearrow}^{G} \cdot p_{k\searrow}^{G}).$$

- Obtain the null distribution of the Ψ_k by computing

$$ {}^{b}\Psi_k^* = -2\log({}^{b}p_{k\nearrow}^{G*} \cdot {}^{b}p_{k\searrow}^{*G}) \qquad b = 1, \ldots, B.$$

- Obtain the p-value for the umbrella alternative on group k as

$$\pi_k = \frac{\#[{}^{b}\Psi_k^* \geq \Psi_k]}{B}.$$

Note that π_k is significant in favor of the umbrella alternative with peak group k. That is, the smaller π_k, the higher the evidence of having an umbrella alternative on group k. In order to evaluate if there is a significant presence of any umbrella alternative, we finally combine the p-values for the umbrella alternative of each group. To do so:

- Obtain the null distribution of the p-value for the umbrella alternative on group k as

$$ {}^{b}\pi_k^* = \frac{\#[\Psi_{\mathbf{k}}^* \geq {}^{b}\Psi_k^*]}{B}, \qquad b = 1, \ldots, B,$$

where $\Psi_{\mathbf{k}}^*$ is the vector with the permutation null distribution of π_k.
- Apply *Tippett's* combining function to the π_k's, providing the observed value of the global test statistic for the umbrella alternative in any group,

$$\Pi = \min(\pi_1, \pi_2, \ldots, \pi_C).$$

Note that small values of Π are significant against the null hypothesis for at least one group.

- Obtain the null distribution of Π by computing

$$^b\Pi^* = \min(^b\pi_1^*, {}^b\pi_2^*, \ldots, {}^b\pi_C^*) \qquad b = 1, \ldots, B.$$

- Obtain the global p-value as

$$\Pi^G = \frac{\#[^b\Pi^* \le \Pi]}{B}.$$

Note that the combining functions are applied simultaneously to each random permutation, providing the null distributions of partial and global tests as well. The NPC methodology applies three times in this testing procedure:

1. when obtaining simple stochastic ordering tests to assess $H_{1k}^{\nearrow}$ and $H_{1k}^{\searrow}$ for the kth group ("direct" combining function);
2. when combining together the partial tests for simple stochastic ordering alternatives, providing a test for the umbrella for each group as if it were the known peak group ("Fisher's" combining function); and
3. when combining together the partial tests for the umbrella on each group ("Tippett's" combining function).

A significant global p-value Π^G indicates that there is evidence in favor of an umbrella alternative. The peak group is then identified by looking at the partial p-values for the umbrella alternatives $\{\pi_1, \pi_2, \ldots, \pi_k, \ldots, \pi_C\}$. The peak group (if any) is then the one with minimum p-value.

The proposed algorithm may still apply with different combining functions at each step. The power behavior of some of them is discussed in Section 5.6. We have run the test for the umbrella alternative on data of Table 5.5. The results are shown in Table 5.6. Here there is a significant presence of an umbrella since the global p-value is equal to 0.014. The group $35-54$ is the peak group since its partial p-value is the smallest one ($p_3 = 0.00299$).

Table 5.6. Permutation test results, WAIS score data. $\Pi^G = 0.014$.

Age	15−19	20−34	35−54	55−69	>70
p_k	0.05794	0.00599	**0.00299**	0.12687	0.94306

5.5.1 The Mack and Wolfe Test

Mack and Wolfe (1981) propose a distribution-free test for umbrella alternatives that is based on a linear combination of (dependent) Mann-Whitney statistics. They first introduce the test statistic in the case of a known umbrella point, and then extend the solution to the case of an unknown peak.

When the peak group is known a priori to be the kth one, then (5.7) is rejected in favor of (5.8) for large values of the test statistic

$$A_k = \sum\sum_{1 \le i < j \le k} U_{ij} + \sum\sum_{k \le i < j \le C} U_{ji}, \qquad (5.10)$$

where, according to previous notation,

$$U_{ij} = \#\{\mathbf{Y_i} < \mathbf{Y_j}\}.$$

That is, U_{ij} is the Mann-Whitney statistic between the ith and jth samples. Let

$$\mu_0(A_k) = \frac{1}{4}\left[N_1^2 + N_2^2 - \sum_{j=1}^{C} n_j^2 - n_k^2 \right],$$

$$\sigma_0^2(A_k) = \frac{1}{72}\left[2(N_1^3 + N_2^3) + 3(N_1^2 + N_2^2) - \sum_{j=1}^{C} n_j^2(2n_j + 3) - n_k^2(2n_k + 3) \right]$$

$$+ \frac{1}{72}[12 n_k N_1 N_2 - 12 n_k^2 N],$$

be the null mean and variance of A_k, where $N_1 = \sum_{j=1}^{k} n_j$, $N_2 = \sum_{j=k}^{C} n_j$, and $N = \sum_{j=1}^{C} n_j$. Then, under (5.7), the asymptotic distribution of $[A_k - \mu_0(A_k)]/\sigma_0(A_k)$ is standard normal. The authors provided the small sample null distribution of (5.10) by expressing A_k as a sum of several independent (under (5.7)) random variables, each of which has an appropriate Mann-Whitney null distribution. Then, by using the properties of the convolution of independent random variables, they were able to provide the null distribution of A_k for a variety of n_j configurations.

The case of an unknown umbrella point is treated similarly, although the authors first need to estimate the unknown umbrella peak k here. Mack and Wolfe proposed to reject (5.7) in favor of the k-unknown alternative (5.8) for large values of the test statistic

$$A_{\hat{k}} = \sum_{k=1}^{C} \chi_k \frac{A_k - \mu_0(A_k)}{\sigma_0(A_k)}. \qquad (5.11)$$

That is, the proposed test statistic is a (weighted) linear combination of peak-known standardized statistics A_j given by (5.10), and $\mu_0(A_j)$ and $\sigma_0(A_j)$ are the corresponding null mean and variance. The random variables $\{\chi_1, \ldots, \chi_C\}$ are indicator variables of which group(s) has been estimated by the observed data to be the peak group(s).

In order to determine the peak group, Mack and Wolfe suggest the following algorithm:

- For $k = 1, \ldots, C$, repeat:
  1. Let

$$Z_k = \sum_{j \neq k}^{C} U_{jk}.$$

Z_k is a two-sample Mann-Whitney statistic computed between the kth sample and the remaining $C - 1$ samples. Large values of Z_k indicate a possible candidate for the peak group.

2. Let

$$\mu_0(Z_k) = \frac{n_k(N - n_k)}{2},$$

$$\sigma_0^2(Z_k) = \frac{n_k(N - n_k)(N + 1)}{12},$$

be the null mean and variance of Z_k.

• Let $Z_{(1)} = \max_k Z_k$ and $r = \#\{Z_k = Z_{(1)}\}$. Then set

$$\chi_k = \begin{cases} 1/r & \text{if } Z_k = Z_{(1)} \\ 0 & \text{otherwise.} \end{cases}$$

With these settings, one can compute the test statistic (5.11), which is an average of known-peak statistics corresponding to the groups with largest Z_k. Note that, within this testing procedure, the probability of having ties in the Z_k's is positive. If the sample sizes are equal, one does not need to compute $\mu_0(Z_k)$ and $\sigma_0^2(Z_k)$ in order to obtain the maximum $Z_{(1)}$. The authors provide critical values corresponding to α sizes of .1, .05, and .01 obtained from exact (for small smaples) and simulated null distribution of $A_{\hat{k}}$. We refer to Mack and Wolfe (1981) for details.

We have implemented an R function performing the Mack and Wolfe test (see Subsection 5.7.2), which provided us the $A_{\hat{k}}$ critical values for the comparative simulation study (see Section 5.6). These values were obtained by running the function on independent 10,000 MC data generations under the null hypothesis of equality in distribution in the C samples. We have only considered balanced designs, although the R function can be applied even with unbalanced samples. The results are reported in Table 5.7.

As regards the example of Section 5.3, the Mack and Wolfe test gives a p-value equal to 0.0328 (through a simulated null distrubtion), which indicates the presence of an umbrella alternative. According to their test, the peak group is the third one (age 35−54) and the value of the test statistic is $A_3 = 2.353$ (see Subsection 5.7.2 and Table 5.8 for details).

5.6 A Comparative Simulation Study

In this section we compare the permutation test introduced in Section 5.5 with Mack and Wolfe's test. The comparison is made through a simulation study under the null hypothesis and under the alternative hypothesis.

Table 5.7. Critical values of the Mack and Wolfe test for some n and C.

		α		
C	n	.10	.05	.01
3	3	1.889	2.324	2.556
3	5	1.837	2.060	2.589
3	7	1.791	2.094	2.609
5	3	1.924	2.228	2.633
5	5	1.914	2.200	2.725
5	7	1.951	2.240	2.776
7	3	1.986	2.233	2.782
7	5	1.980	2.285	2.831
7	7	1.997	2.276	2.834

Table 5.8. Mack and Wolf results, WAIS score data.

Age	$15 - 19$	$20 - 34$	$35 - 54$	$55 - 69$	> 70
$A_{\hat{k}}$	1.114013	2.118296	2.353394	0.6657503	-1.114013
Z_k	-0.5773503	1.010363	**1.876388**	-0.2886751	-2.020726

$$1 - F_{A_{\hat{k}}}(2.353394) = 0.0328$$

The chosen settings are $K = 5$ groups with $n_j \equiv 3$ observations each ($j = 1, \ldots, 5$). The simulated data have a standard normal distribution, possibly with some nonrandom location shifts in some groups (under the alternative hypothesis). Each simulation is based on 1000 independent Monte Carlo data generations.

Let us discuss the results in Table 5.9 first. Here the rejection rates of the null hypothesis at different α sizes are reported. The location shifts (δ_k's) are reported at the top of the table. The first part of Table 5.9 refers to the permutation test. Note how the rejection rates in the global test column (indicated by "Π^G") are very close to the nominal ones. Then, for each group, the rejection rates of the partial tests for peak-known umbrellas (the π_k's) are also shown. In order to account for multiplicity, the partial p-values are compared with the adjusted α level through a Bonferroni correction (therefore the nominal level for the partial tests is $\alpha/5$). Finally, the probability of observing a peak in group k conditional on the rejection of the global null hypothesis is reported for the same α sizes. The second part of Table 5.9 refers to Mack and Wolfe's test. Here the p-value is reported in the "p" column, and the rejection rates of the null hypothesis are still very close to the nominal ones. Then, the probability of observing a maximum in group $k = 1, \ldots, 5$ conditional on the rejection of the null hypothesis is reported for each group. These probabilities should be close to $1/5$ under H_0.

Table 5.10 shows the behavior of the two tests under an umbrella alternative with peak on the third group. The table structure is the same as before.

Table 5.9. Rejection rates of permutation and Mack and Wolfe tests under H_0.

	δ_1	δ_2	δ_3	δ_4	δ_5	
	0	0	0	0	0	
			Permutation test			
α	p_1	p_2	p_3	p_4	p_5	Π^G
0.05	0.015	0.012	0.015	0.009	0.024	**0.050**
0.1	0.023	0.016	0.039	0.008	0.021	**0.096**
0.2	0.048	0.041	0.053	0.018	0.045	**0.175**
α	$P\{\pi_k = \min_j \pi_j \mid \Pi^G \leq \alpha\}$					Total
0.05	0.111	0.333	0.111	0.333	0.111	1
0.1	0.238	0.143	0.095	0.238	0.286	1
0.2	0.278	0.194	0.111	0.222	0.194	1
		Mack & Wolfe's test				
α	$P\{k = \max_j \pi_j \mid p \leq \alpha\}$					p
0.05	0.143	0.161	0.304	0.250	0.143	**0.056**
0.1	0.173	0.163	0.288	0.202	0.173	**0.104**
0.2	0.137	0.235	0.225	0.209	0.194	**0.211**

Note that the rejection rates of the global test are far bigger than the nominal levels, but the permutation test is more powerful than the competitor. As far as the permutation test is concerned, the rejection rates of the partial tests (accounting for multiplicity) are directly proportional to the size of the δ_k's, and group 3 has been detected as the peak group about 50% of the times that the global null hypothesis has been rejected at all α levels (however, δ_k sizes are modest compared with the variance of data distribution $\sigma^2 = 1$). The Mack and Wolfe test shows a good performance, too, and here the probability of detecting the third group as the peak group seems more stable.

In Table 5.11, we have set the location shifts in order to simulate an anti umbrella alternative. That is, data are not under the null hypothesis, but the true alternative hypothesis is not of the umbrella kind since the trend is first decreasing and then increasing. Neither of the tests should recognize this kind of alternative. Indeed, both the permutation test global p-value Π^G and Mack and Wolfe p-value are smaller than the related nominal levels. Nevertheless, the third group (where the trend has its minimum) is never recognized as the peak group by the permutation test, whereas Mack and Wolfe's test has a positive probability of observing the third group as a maximum.

In the proposed algorithm, several combining functions are applied. Which combining function is to be applied depends on the problem we are dealing with: Some combining functions are more sensitive than others to specific alternatives. For instance, if the alternative hypothesis is "half-umbrella" (that is, no trend up to the $\hat{k}$th group, then a descreasing trend),

Table 5.10. Rejection rates of permutation and Mack and Wolfe tests under the umbrella alternative.

	δ_1	δ_2	δ_3	δ_4	δ_5	
	0	0.5	1	0.5	0	

			Permutation test			
α	p_1	p_2	p_3	p_4	p_5	Π^G
0.05	0.021	0.073	0.127	0.031	0.007	**0.243**
0.1	0.022	0.101	0.220	0.052	0.008	**0.359**
0.2	0.037	0.152	0.331	0.123	0.033	**0.547**

α	$P\{\pi_k = \min_j \pi_j \mid \Pi^G \leq \alpha\}$					Total
0.05	0.000	0.115	0.692	0.192	0.000	1
0.1	0.022	0.156	0.556	0.244	0.022	1
0.2	0.045	0.212	0.485	0.212	0.045	1

			Mack & Wolfe's test			
α	$P\{k = \max_j \pi_j \mid p \leq \alpha\}$					p
0.05	0.011	0.167	0.649	0.167	0.007	0.228
0.1	0.019	0.176	0.620	0.169	0.017	0.347
0.2	0.018	0.184	0.598	0.180	0.020	0.505

Table 5.11. Rejection rates of permutation and Mack and Wolfe tests under the anti umbrella alternative.

	δ_1	δ_2	δ_3	δ_4	δ_5	
	1	0.5	0	0.5	1	

			Permutation test			
α	p_1	p_2	p_3	p_4	p_5	Π^G
0.05	0.014	0.004	0.010	0.010	0.015	**0.014**
0.1	0.020	0.009	0.004	0.007	0.032	**0.055**
0.2	0.027	0.013	0.006	0.017	0.042	**0.091**

α	$P\{\pi_k = \min_j \pi_j \mid \Pi^G \leq \alpha\}$					Total
0.05	0.000	0.000	0.000	0.000	1.000	1
0.1	0.300	0.100	0.000	0.100	0.500	1
0.2	0.294	0.059	0.000	0.118	0.529	1

			Mack & Wolfe's test			
α	$P\{k = \max_j \pi_j \mid p \leq \alpha\}$					p
0.05	0.417	0.083	0.000	0.000	0.500	0.012
0.1	0.371	0.057	0.029	0.029	0.514	0.041
0.2	0.347	0.056	0.042	0.083	0.472	0.087

$$H_1 : \left\{ \bigcap_{j=1}^{\hat{k}} H_{0(j)} \right\} \bigcup \left\{ \bigcup_{j=\hat{k}}^{K} H_{1(j)}^{\searrow} \right\},$$

then Tippett's function combining the partial tests for simple stochastic ordering alternatives ($H_{1(k)}^{\nearrow}$ and $H_{1(k)}^{\searrow}$) would be more sensitive for this specific alternative, as it requires at least one significant argument in order to produce large values of the observed global test statistic. For the umbrella alternatives, we suggest applying Fisher's combining function, which is generally more powerful when some partial tests are under the alternative. In particular, we might change the combining functions at steps (2) and (3) of the procedure, as indicated at the end of the previous section.

We have considered two scenarios: one real umbrella and one "half umbrella" alternative when $K = 5$ and $n_j = 3$. The first battery of simulations has been obtained by applying Fisher's combining function at step 2 and then by applying Fisher's, Liptak's, and Tippett's combining function at step 3. The second battery of simulations has been obtained by applying Tippett's combining function at step 2 and Fisher's, Liptak's, and Tippett's at step 3. The power results of the global test for umbrella alternatives are reported in Table 5.12. For each scenario (real umbrella $\nearrow\searrow$ or half-umbrella $\longrightarrow\searrow$ alternative) and each sequence of combining functions, the rejection rates of the global null hypothesis are reported. Clearly, when there is a real umbrella alternative, the proposed test is usually more powerful than when the alternative is the half-umbrella setting. In our opinion, the best choice is to apply Fisher's combining function at step 2 and then apply Tippett's combining function at step 3. The half-umbrella alternative settings were $\delta_1 = \delta_2 = \delta_3 = 1$, $\delta_4 = 0.5$, and $\delta_5 = 0$ with standard normal errors.

Table 5.12. Rejection rates of the permutation test with $\alpha = 0.05$ under different scenarios.

	Alternative			
	$\nearrow\searrow$	$\longrightarrow\searrow$	$\nearrow\searrow$	$\longrightarrow\searrow$
Step 3 ψ	Step 2: Fisher's ψ		Step 2: Tippett's ψ	
Fisher's	0.332	0.216	0.177	0.294
Liptak's	0.338	0.179	0.284	0.256
Tippett's	0.246	0.262	0.213	0.196

5.7 Applications with R

In this section, we will see in detail the topics of this chapter within the R environment.

5.7.1 One-Way ANOVA with R

The `aov_perm.r` function performs the one-way ANOVA test as described in Section 5.2. It has been thought to perform *pooled* permutations when the design is unbalanced and *synchronized* permutations when the design is balanced (default is `balanced = TRUE`). Here the minimum achievable significance levels of the global test are multiples of (see Section 6.3)

$$\frac{1}{2}\binom{2n}{n}.$$

Partial tests on pairwise comparisons and a global test assessing significance against the null hypothesis $F_1(y) \overset{d}{=} F_2(y) \cdots \overset{d}{=} F_C(y)$ are done simultaneously. We can refer to the example of table 5.5 to run the one-way ANOVA with multiple comparisons. Type

```
> source("aov_perm.r")
> C<-5
> n<-3
> x<-c(8.62,9.94,10.06,9.85,10.43,11.31,9.98,10.69,11.40,
+ 9.12, 9.89,10.57,4.80,9.18,9.27)
> y<-rep(seq(1,C),each=n)
> set.seed(11)
> aov.perm(x,y,B=10000)
$Global.p.value
[1] 0.1007

$Partial.p.value
      Diff p sig.
1-2 -0.99 0.2973
1-3 -1.15 0.1990
1-4 -0.32 0.6975
1-5  1.79 0.2995
2-3 -0.16 0.6944
2-4  0.67 0.5000
2-5  2.78 0.1007
3-4  0.83 0.1973
3-5  2.94 0.1007
4-5  2.11 0.2995
```

Note that the `aov.perm` function obtains the Monte Carlo distribution of the global test statistic, and therefore the global *p*-value is approximated (in this case, the exact global *p*-value is equal to $1/10$ and all partial *p*-values are multiples of $1/10$). If *pooled* permutations are to be applied, type

```
> set.seed(11)
> aov.perm(x,y,B=10000,balanced=FALSE)
```

```
$Global.p.value
[1] 0.055

$Partial.p.value
     Diff p sig.
1-2 -0.99 0.4276
1-3 -1.15 0.3828
1-4 -0.32 0.7807
1-5  1.79 0.1937
2-3 -0.16 0.8941
2-4  0.67 0.5809
2-5  2.78 0.0184     *
3-4  0.83 0.5036
3-5  2.94 0.0065    **
4-5  2.11 0.1189
```

The global test is moderately significant against the null hypothesis (Global.-p.value). The difference in the results is due to the higher cardinality of the support of the test statistics when polled permutations are applied. There seems to be higher evidence against the null hypothesis with pooled permutations, but note that the synchronized permutation permutations global p-value was at the minimum attainable significance level. The output of this function also includes the pariwise-comparison table (Partial.p.value). The labels of the rows indicate which comparisons are considered, the Diff. column reports the observed mean differences between the groups considered, and the last column reports the p-values of the partial tests (*not* adjusted for multiplicity). As a final help, the partial p-values are highlighted with the usual R convention:

```
Signif. codes:  0 *** 0.001 ** 0.01 * 0.05 . 0.1   1
```

The parametric F test and Kruskal-Wallis test give the following results:

```
> t<-aov(x~factor(y))
> summary(t)
            Df  Sum Sq Mean Sq F value Pr(>F)
factor(y)    4 16.5580  4.1395  2.3686 0.1225
Residuals   10 17.4762  1.7476

> TukeyHSD(t)
$'factor(y)'
     diff       lwr      upr     p adj
2-1  0.99 -2.562357 4.542357 0.8839618
3-1  1.15 -2.402357 4.702357 0.8196890
4-1  0.32 -3.232357 3.872357 0.9980160
5-1 -1.79 -5.342357 1.762357 0.4971801
3-2  0.16 -3.392357 3.712357 0.9998706
```

```
4-2 -0.67 -4.222357 2.882357 0.9683148
5-2 -2.78 -6.332357 0.772357 0.1489661
4-3 -0.83 -4.382357 2.722357 0.9339450
5-3 -2.94 -6.492357 0.612357 0.1194350
5-4 -2.11 -5.662357 1.442357 0.3511889


> kruskal.test(x,y)



        Kruskal-Wallis rank sum test

data:  x and y
Kruskal-Wallis chi-squared = 7.2333, df = 4, $p$-value = 0.1241
```

5.7.2 Umbrella Alternatives with R

Let us introduce the Mack and Wolfe test first. In order to obtain a p-value, one needs to simulate the null distribution of the test statistic. Alternatively, the acceptance/rejection rule can be derived from the value of the test statistic and the critical values reported in Mack and Wolfe (1981). We are going to show in detail how the results of Table 5.8 have been obtained. First, load the Mack and Wolfe function in the R environment, input the data of Table 5.5 (in the vector x) and the labels of the groups (in the vector y), and then run the function as follows:

```
> source("MW.r")
>   C<-5
>   n<-3
>   x<-c(8.62,9.94,10.06,9.85,10.43,11.31,9.98,10.69,11.40,
+ 9.12, 9.89,10.57,4.80,9.18,9.27)
> y<-rep(seq(1,C),each=n)
> y
 [1] 1 1 1 2 2 2 3 3 3 4 4 4 5 5 5
> MW(x,y)
$T
[1] 2.353394

$A
          [,1]
[1,]   1.1140133
[2,]   2.1182964
[3,]   2.3533936
[4,]   0.6657503
[5,]  -1.1140133
```

```
$Z
           [,1]
[1,]  -0.5773503
[2,]   1.0103630
[3,]   1.8763884
[4,]  -0.2886751
[5,]  -2.0207259


$peak
[1] 3
```

The output of the MW function is a list containing the following objects:

- Z: the standardized Z_k statistics to estimate the peak group;
- A: the standardized A_k statistics for the known peak;
- T: the observed value of the Mack and Wolfe test statistic $A_{\hat{k}}$;
- peak: the estimated peak group.

In the example, $C = 5$, $n_k \equiv 3$, and the $A_{\hat{k}}$ statistic is equal to 2.353394. According to both the critical values of Table 5.7 and Mack and Wolfe (1981), the test is significant at a 5% level (the critical values are respectively equal to 2.228 and 2.239 for $\alpha = .05$). If a p-value is desired, instead one can simulate the null distribution by running the MW.r function a large number of times with data under H_0 and by storing each time the value of the test statistic T. We can do this, for instance, by simulating standard normal data (data from any continuous distribution with finite first and second moments will be suitable as well). For instance, if we consider a simulation with 10,000 Monte Carlo normal data generations,

```
>   set.seed(1)
>   MC<-10000
>   T<-array(0,dim=c(MC,1))
>   for(cc in 1:MC){
+   z<-rnorm(n*C)
+   T[cc]<-MW(z,y)$T
+   }
>   p<-sum(T>=2.353394)/MC
>   p
[1] 0.036
```

After a while, the simulated null distribution of $A_{\hat{k}}$ is stored in the vector T, and the approximated p-value of the test is equal to 0.036.

The permutation test is performed by the umbrella.r function, which requires the same input of the MW.r function. The number of permutations considered is set equal to 1000 by default, but it can be specified as input:

```
> source("combine.r")
> source("t2p.r")
```

```
> source("umbrella.r")
> set.seed(6)
> umbrella(x,y,B=10000)
$Global.p.value
[1] 0.0264

$Partial.p.values
[1] 0.07099290 0.00669933 0.00549945 0.11468853 0.92910709

$Max
[1] 3
```

We have just run the testing procedure with $B = 10,000$ permutations. There are two warning messages produced by the combining function `combine.r`. This function allows us to actuate the nonparametric combination of partial p-values/test statistics, depending on the input and the specified combining function. As pointed out in Section 5.5, Fisher's combining function is applied at step 2 and Tippett's combining function is applied at step 3 of the algorithm. The global p-value is equal to 0.0264, which is even more significant than Mack and Wolfe's p-value. Since the global test is significant, we can estimate the peak group by looking at the partial p-values. The minimum partial p-value is the one related to group 3, which is highlighted in the `Max` output. It is worth noting that the `umbrella.r` function (like the `MW.r` function) *always* gives an estimated maximum, but the global p-value (respectively, the $A_{\hat{k}}$ statistic) has to be taken into account to determine if there is a significant umbrella trend or not. For instance, we can repeat the tests with simulated data from a uniform distribution:

```
>   set.seed(10)
>   x<-runif(n*C)
>   umbrella(x,y)
$Global.p.value
[1] 0.528

$Partial.p.values
[1] 0.5754246 0.7402597 0.5774226 0.1538462 0.4255744

$Max
[1] 4

>
>   MW(x,y)
$T
[1] 1.028887

$A
```

```
                  [,1]
[1,]  -0.3038218
[2,]  -0.5447048
[3,]  -0.1307441
[4,]   1.0288868
[5,]   0.3038218

$Z
                  [,1]
[1,]   0.0000000
[2,]  -0.7216878
[3,]  -0.2886751
[4,]   1.4433757
[5,]  -0.4330127


$peak
[1] 4

> sum(T>=1.028887)/MC
[1] 0.4723
```

Since the global p-value of the permutation test and the approximated p-value of Mack and Wolfe's test are both not significant, both tests agree that there is not a significant umbrella trend within these data (therefore there is no significant peak or maximum). Finally, Figure 5.4 was obtained as follows:

```
>  x<-c(8.62,9.94,10.06,9.85,10.43,11.31,9.98,10.69,11.40,
+ 9.12, 9.89,10.57,4.80,9.18,9.27)
> boxplot(x~y)
> m<-array(0,dim=c(C,1))
> for(cc in 1:C){
+ m[cc]<-mean(x[y==cc])
+ points(cc,m[cc],pch=16)
+ }
> lines(m,lty="dotted")
```

6

Synchronized Permutation Tests in Two-way ANOVA

6.1 Introduction

In experimental planning, factorial experiments are of particular interest, as they allow us to separately examine the effects of two or more factors in all their possible combinations. In the usual linear model for the analysis of variance, if the error components are not normally distributed, parametric analysis may not be appropriate, even if Rasch and Guiard (2004), recalling previous results by Ito (1969), show that parametric tests are generally robust also in the presence of some nonnormal distributions. When we wish to apply nonparametric tests to the two-way ANOVA layout, problems arise with the exchangeability of the response, which is not satisfied since observations with different treatments might have different expected values. To cope with this problem, either a restricted kind of randomization or the use of residuals is required in order to obtain separate tests for main factors and interaction. In this chapter, based on the concept of *synchronized permutations*, we introduce an exact permutation solution (Pesarin, 2001; Salmaso, 2003; Basso et al., 2007) for testing for fixed effects in replicated two-way factorial designs with continuous responses. This permutation solution, since it is conditional on a set of sufficient statistics, is a distribution-free nonparametric test. It is worth noting that asymptotically distribution-free (but not exact) tests could also be developed using the approach by Draper (1988) or the recent development of a generalization of the Kruskal-Wallis approach to two- and three-way layouts given in Brunner and Puri (2001). Among the exact tests, we can differentiate synchronized permutation tests from those inspired by the two-way ANOVA F test, such as the tests proposed by Edgington (1995), Maritz (1995), and Sprent (1998). For the class of approximated two-way ANOVA permutation tests, we mention those proposed by Manly (1997) and Anderson (2001), which are based on some kind of F statistic. The remaining approximated permutation tests proposed in the literature are inspired by multiple regression analysis, such as that proposed by Cade and Richards (1996) and one of those discussed by Kennedy and Cade (1996). Finally, there

D. Basso et al., *Permutation Tests for Stochastic Ordering and ANOVA*, Lecture Notes in Statistics, 194, DOI 10.1007/978-0-387-85956-9_6,
© Springer Science+Business Media, LLC 2009

are other unrestricted permutation procedures, also belonging to the field of multiple regression analysis, that permute not the observations but the residuals and are calculated by means of an estimate of regression models (see also Anderson and ter Braak, 2003).

The power of the proposed permutation solution is comparable with that of the parametric solution when the latter is applicable. Let us remember that permutation tests are conditional procedures in which conditioning is with respect to a set of joint sufficient statistics under the null hypothesis. Hence, the permutation approach for a two-way layout should be based on such a set of joint sufficient statistics. We will present the theory in the case of a balanced two-factor design where factor A has I levels and factor B has J levels. The linear model is

$$Y_{ijk} = \mu + \alpha_i + \beta_j + \alpha\beta_{ij} + \varepsilon_{ijk} \qquad \begin{cases} i = 1, \ldots, I, \\ j = 1, \ldots, J, \\ k = 1, \ldots, n, \end{cases}$$

where Y_{ijk} are the experimental responses, μ is the general mean, α_i and β_j are the main factor effects, $\alpha\beta_{ij}$ are interaction effects, ε_{ijk} are exchangeable experimental errors, with zero mean, from an unknown continuous distribution P and n is the number of replicates in each cell. The usual side conditions are

$$\sum_i \alpha_i = 0, \ \sum_j \beta_j = 0, \ \sum_i \alpha\beta_{ij} = 0 \, \forall j, \ \sum_j \alpha\beta_{ij} = 0 \, \forall i \qquad (6.1)$$

Usually, the experimenter's major interest is testing separately for two main effects and interactions. Hence, there are three sub null hypotheses of interest − $H_{0A} : \alpha_i = 0 \forall i$, $H_{0B} : \beta_j = 0 \forall j$, $H_{0AB} : \alpha\beta_{ij} = 0 \forall i, j$ − and the emphasis is on finding three separate and possibly uncorrelated tests. What experimenters are generally looking for is, for instance, to test H_{0A} against $H_{1A} : \exists i : \alpha_i \neq 0$, irrespective of whether H_{0B} or H_{0AB} is true or not, etc. In order to attain this goal within a permutation framework, we must find the proper set of jointly sufficient statistics for all three testing sub problems: H_{0A} irrespective of $H_{0B} \cup H_{0AB}$ is true or not, H_{0B} irrespective of $H_{0A} \cup H_{0AB}$ is true or not, and H_{0AB} irrespective of $H_{0A} \cup H_{0B}$ is true or not. In this framework, this set is $\mathbf{y} = [\mathbf{y}_{11}, \mathbf{y}_{12}, \ldots, \mathbf{y}_{IJ}]'$, the vector of the observed response partitioned into $I \times J$ blocks. This is due to the definition of sufficient statistic: in fact, two points of the sample space, $\mathbf{y}$ and $\mathbf{y}'$, lay in the same orbit of a sufficient statistic if the likelihood ratio

$$\frac{L(\boldsymbol{\alpha}, \boldsymbol{\beta}, \boldsymbol{\alpha\beta}; \mathbf{y})}{L(\boldsymbol{\alpha}, \boldsymbol{\beta}, \boldsymbol{\alpha\beta}; \mathbf{y}')} = \frac{\prod_{ijk} L(\alpha_i, \beta_j, \alpha\beta_{ij}; y_{ijk})}{\prod_{ijk} L(\alpha_i, \beta_j, \alpha\beta_{ij}; y'_{ijk})} = h(\mathbf{y}, \mathbf{y}')$$

does not depend on the parameters α_i, β_j, $\alpha\beta_{ij}$. This only occurs if $\mathbf{y}'$ is a permutation of the units within the blocks of $\mathbf{y}$. From a naive point of view, we are only allowed to permute data within each block, but this is useless since any permutation within blocks gives the same value of any suitable test statistic, and hence no real distribution for the test statistic can be found. Thus, we must look either for approximate solutions or for a special kind of restricted permutation strategy.

6.2 The Test Statistics

Let us consider the permutation structure of an intermediate statistic to separately compare the factor A effects α_i and α_s at level j of factor B

$$^aT_{is|j} = \sum_k y_{ijk} - \sum_k y_{sjk}. \tag{6.2}$$

For each level of factor B, there are $I(I-1)/2$ such intermediate statistics, each comparing two different effects of factor A. Now suppose that ν units are randomly exchanged between blocks A_iB_j and A_sB_j, ν being invariant with respect to the levels of factor B. With obvious notation, the permutation structure of these intermediate statistics is

$$^aT_{is|j}^* = (n - 2\nu)[\alpha_i - \alpha_s + \alpha\beta_{ij} - \alpha\beta_{sj}] + \bar{\varepsilon}_{is|j}^*, \tag{6.3}$$

where the symbol "*" denotes that we have performed a random synchronized permutation (since ν is invariant w.r.t. the levels of factor B) among units in blocks A_iB_j and A_sB_j, $\bar{\varepsilon}_{is|j}^* = (n-\nu)[\bar{\varepsilon}_{ij}^* - \bar{\varepsilon}_{sj}^*] + \nu[\bar{\varepsilon}_{ij}^* - \bar{\varepsilon}_{sj}^*]$ is the permutation error term, and $\bar{\varepsilon}_{ij}^*, \bar{\varepsilon}_{sj}^*, \bar{\varepsilon}_{ij}^*, \bar{\varepsilon}_{sj}^*$ are random sampling means of $n-\nu$ and ν errors, respectively, from different pairs of blocks. Note that $^aT_{is|j}^*$ depends either on the effects of factor A or on interaction effects, and hence we need a further step in order to separately test for main factors and interaction. According to the side conditions $\sum_i \alpha\beta_{ij} = 0 \,\forall\, j$, $\sum_j \alpha\beta_{ij} = 0 \,\forall\, i$, the sum

$$\sum_j {}^aT_{is|j}^*, \tag{6.4}$$

whose permutation structure is $J(n - 2\nu)[\alpha_i - \alpha_s] + \sum_j \bar{\varepsilon}_{is|j}^*$, only depends on effects α_i and α_s of factor A and on a linear combination of errors. Hence it allows for an exact permutation test for A, irrespective of whether other factor effects are present or not. This test is then

$$^aT_A^* = \sum_{i<s} \left[\sum_j {}^aT_{is|j}^* \right]^2, \tag{6.5}$$

where the inner sum is squared to avoid the vanishing of some effects of factor A and the outer sum is made over all possible pairs of levels $1 \leq i < s \leq I$. Moreover, to test for interaction effects, let us first consider the permutation structure of the difference of two intermediate statistics at two distinct levels of factor A and factor B:

$$^aT^*_{is|j} - ^a T^*_{is|h} = (n - 2\nu)[\alpha\beta_{ij} - \alpha\beta_{ih} - \alpha\beta_{sj} + \alpha\beta_{sh}], +\bar{\varepsilon}^*_{is|j} - \bar{\varepsilon}^*_{is|h}.$$

The elements of this difference depend only on the effects of interaction AB and on a random permutation of errors. Thus a separate test for interaction is

$$^aT^*_{AB} = \sum_{i<s}\sum_{j<h} [^aT^*_{is|j} - ^aT^*_{is|h}]^2, \tag{6.6}$$

where the superscript "a" in the test statistic specifies that the test for interaction has been obtained from intermediate statistics for factor A. In order to test separately for H_{0B} and H_{0AB}, we proceed in accordance with the previous strategies. Hence, by interchanging indices, we have respectively

$$^bT^*_B = \sum_{j<h}\left[\sum_i {}^bT^*_{jh|i}\right]^2, \tag{6.7}$$

and

$$^bT^*_{AB} = \sum_{j<h}\sum_{i<s} [^bT^*_{jh|i} - ^bT^*_{jh|s}]^2, \tag{6.8}$$

where $^bT^*_{jh|i} = \sum_k y^*_{ijk} - \sum_k y^*_{ihk}$; i.e., the intermediate statistic to separately compare the factor B effects β_j and β_h is obtained by exchanging the same number ν of units between blocks A_iB_j and A_iB_h, $i = 1, \ldots, I$. For each level of factor A, there are $J(J-1)/2$ such intermediate statistics. Note that $^aT^*_{AB}$ is obtained from synchronized permutations involving the row factor A, whereas $^bT^*_{AB}$ is obtained from permutations involving the column factor B. Each statistic for interaction only depends on interaction effects and exchangeable errors, and so they are jointly and equally informative. Thus, their linear combination $T^{**}_{AB} = ^aT^*_{AB} + ^bT^*_{AB}$ is a separate exact test for interaction. Once we have defined the test statistics for main effects and interaction, we need to define the synchronized permutation strategy in order to apply these tests.

6.3 Constrained and Unconstrained Synchronized Permutations

The basic concept of synchronized permutations is exchanging the same number ν of units within each pair of blocks considered. There are two ways to obtain a synchronized permutation: exchanging units in the same original positions within each block (constrained synchronized permutations, CSPs) or exchanging units without considering their original position (unconstrained synchronized permutations, USPs). The core point is that in both cases the same number of units must be exchanged within all couples of cells in the pairs of rows or columns considered. We implemented two distinct algorithms, one for CSPs and another for USPs, since they differ both in the probability of observing a single synchronized permutation and in the cardinality of distinct permutation test statistic values. Let us refer to constrained synchronized permutations first: Since these are defined by the number of units being exchanged (ν) at their original positions within each block, it is easy to see that the total number of CSPs depends only on which exchange has been made in the first pair of blocks. Since there are

$$C_{CSP} = \binom{2n}{n}$$

possible ways to exchange units in the first pair of blocks, C_{CSP} is the cardinality of the CSPs. Another point to take into account is the cardinality of distinct permutation test statistics (e.g., the number of distinct $^aT_A^*$'s if we are testing for factor A): the squaring operators in formulas (6.5), (6.7), (6.6), and (6.8) produce a symmetry (i.e., there are two distinct permutations generating the same value of $^aT_A^*$, and hence the total number of distinct $^aT_A^*$'s is $C_{CSP}/2$). Here and in what follows we will only refer to the test for factor A, as the same strategy for factor B can be easily obtained in the same way and the tests for interaction are derived from those of the main factors. A way to obtain Monte Carlo CSPs is to consider, for each pair of distinct levels of factor A, the matrix made of $2n$ rows and J columns corresponding to the levels of factor B. Each vector of this matrix is the pooled vector of the observations from blocks A_iB_j and A_sB_j. We consider a random permutation of the $2n$ units of the first vector and apply the same permutation to the remaining $J - 1$ vectors. Then each vector is partitioned in two blocks of n observations each. This way of proceeding guarantees that the same number of exchanges between each pair of blocks has been made. It also guarantees that each distinct value of $^aT_A^*$ is equally likely. This is important since, in a permutation approach, each value of $^aT_A^*$ has the same probability in the null distribution. If n is relatively small, it is also possible to consider *all* possible distinct CSPs, thus obtaining the exact distribution of $^aT_A^*$.

The main difference between CSPs and USPs is that USPs do not require the exchanged units to be in the same original position within the blocks. Thus, from a naive point of view, we could apply the same algorithm with an initial random shuffling within each single block in order to obtain the USPs. Unfortunately, this way of proceeding does not guarantee that the test statistic follows a (discrete) uniform distribution, as we will describe with a simple example. Consider the 2×2 full factorial design with $n = 2$. The number of distinct CSPs is 6 and the number of distinct values of $^aT_A^*$'s is 3, so each one has a probability of $1/3$. In particular, one value is obtained when $\nu = 0$, and the same value is also obtained when $\nu = 2$. The remaining two distinct values are obtained when $\nu = 1$. Now there are $(2!)^4 = 16$ ways of shuffling units independently within each block. Note that if $\nu = 0$ or $\nu = 2$, the initial shuffling does not change the value of $^aT_A^*$; hence the probability of observing

$^aT_A^* = {}^aT_A$ is still $1/3$. This does not happen when $\nu = 1$ because the initial shuffling makes $^aT_A^*$ take distinct values at each distinct combination of labels. There are 16 such distinct combinations, but the symmetry in the test statistic still holds, and therefore the cardinality of distinct values when $\nu = 1$ is 8. These values have the same probability given $\nu = 1$, so here the probability of obtaining a single value of the test statistic is $P[^aT_A^* = {}^at_A^*, \nu = 1] = (2/3)/8 = 1/12$. This simple example shows that we have to apply another algorithm in order to build the USPs since in this example the observed value of the test statistic has a probability 4 times bigger than the other values, and this would inflate the p-value.

Instead, we would need the nine distinct values to have the same probability of $1/9$. We can investigate the cardinality of USPs as follows. Adapting USPs, units being exchanged within each pair of blocks may differ from the original positions of units exchanged in the first pair of blocks. The only requirement is that the number of exchanges is the same. Hence, for any couple of levels of factor A and a given number of exchanges ν, there are

$$\binom{n}{\nu}^{2J}$$

possible ways to choose the same number of units to be exchanged in the $2J$ cells. Since the number of possible couples of levels of factor A is $I(I-1)/2$, the total number of USPs when testing for factor A is

$$\sum_{\nu=0}^{n} \binom{n}{\nu}^{J \times I(I-1)}$$

Things get harder if we wish to calculate the total number of distinct values of $^aT_A^*$. The symmetry in the test statistic plays a different role when n is odd or even. Let us consider the case when n is odd first. It is easy to see that we can obtain the same values of $^aT_A^*$ when $\nu = x$ or $\nu = n - x$, $0 \le x \le (n-1)/2$. Hence, the cardinality of distinct values of $^aT_A^*$ when n is odd is

$$C_{USP}^o = \sum_{\nu=0}^{(n-1)/2} \binom{n}{\nu}^{J \times I(I-1)}. \tag{6.9}$$

and we only need to consider $\nu = x$ exchanges in order to obtain all possible values of $^aT_A^*$ by applying an initial shuffling of units within each block. The initial shuffling within each cell guarantees the values of $^aT_A^*$ obtained from a given number of exchanges to be equally likely, i.e., given ν, we have

$$P[^aT_A^* = {}^at_A^* | \nu] = \frac{1}{\binom{n}{\nu}^{J \times I(I-1)}}. \tag{6.10}$$

Furthermore, the probability of making ν exchanges is

$$P[N = \nu] = \frac{\binom{n}{\nu}^{J \times I(I-1)}}{\sum_{\pi=0}^{(n-1)/2} \binom{n}{\pi}^{J \times I(I-1)}}. \qquad (6.11)$$

Thus, a two-step algorithm can guarantee the values of the test statistic to be equally likely by first choosing the number of exchanges to be made in accordance with (6.11), then shuffling the units within each cell, and finally exchanging the first ν units between each pair of cells. The same strategy holds when n is even, with some care. Here we obtain the same values of $^aT_A^*$ when $\nu = x$ or $\nu = n - x, 0 \leq x < n/2$, because of the symmetry. Instead, when $\nu = n/2$, each distinct value of $^aT_A^*$ is repeated twice within a couple of distinct levels of factor A since for a given shuffling we get the same value of $^aT_A^*$ by exchanging the first $n/2$ units or the last $n/2$ units within a pair of cells. Hence the total number of distinct values of $^aT_A^*$ when n is even becomes

$$C_{USP}^e = \left[\sum_{\nu=0}^{n/2-1} \binom{n}{\nu}^{J \times I(I-1)} + \left[\frac{1}{2} \binom{n}{n/2} \right]^{2J} \right]^{I(I-1)/2}, \qquad (6.12)$$

and we can apply the same strategy as before by choosing the number of exchanges to be made in accordance with (6.12). The cardinality of distinct values of $^aT_A^*$ rapidly increases with n, I, and J, so we recommend using USPs when few replicates are available (say $n \leq 3$). If the number of replicates is greater than 3, one can easily apply the CSPs. Of course, a different choice between CSPs and USPs affects the minimum achievable significance level. Being a permutation test, the attainable significance levels are multiples of $1/C$, where C is the cardinality of distinct values of the test statistic. Hence, if n is too small, CSPs give a minimum achieved significance level that is higher than the type I error rates commonly used. Figure 6.1 clearly represents this situation: It concerns a simulation under H_{0A} for a 2×3 replicated factorial design with $n = 3, 5, 10$, respectively. The lines are the cumulative distribution functions of the p-values obtained by applying CSP and/or USP tests to the effects of factor A. It is clear (see the picture on the left of Figure 6.1) that applying the CSPs when $n = 3$ gives a discrete uniform distribution function (solid line) for the p-values of factor A. In particular, the minimum achievable significance level is $2/\binom{6}{3} = 1/10$. However, when USPs are applied and $n = 3$ (dotted line), there is a sufficiently high number of distinct $^aT_A^*$'s ($C_{USP}^o = 730$) to guarantee a nearly continuous uniform distribution for the p-value of A when H_{0A} is true, and hence the usual nominal significance levels can be used. The difference between CSPs and USPs rapidly decreases with growing n (see the second and third pictures of Figure 6.1). For instance, if $n = 2, 3, 4, 5$, then C_{CSP} is $3, 10, 35, 126$, respectively. The graphs in Figure 6.1 are obtained by 1000 independent data generations. In the cases of three and five replicates, the CSP distribution of the test statistic is exact (i.e., we performed all possible distinct constrained synchronized permutations, as this leads to a cardinality of the support equal to 10 and 126, respectively). For USPs or when $n = 10$, 1000 Monte Carlo synchronized permutations have been applied.

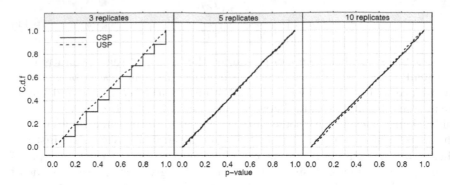

Fig. 6.1. Cumulative distribution function under H_0 of the p-values obtained from CSPs and USPs for a factor with two levels in a 2×3 design with $n = 3, 5, 10$ replicates.

6.4 Properties of the Synchronized Permutation Test Statistics

In this section, we give some formal proofs about the properties of the synchronized permutation tests, such as uncorrelatedness among the test statistics and consistency. The unbiasedness of the test statistic will only be proved for $I = J = 2$ since in that case the squaring operator in the test statistics can be removed so $^aT_A^*$, $^bT_B^*$, $^aT_{AB}^*$, and $^bT_{AB}^*$ have symmetric distributions. It does not seem possible to prove unbiasedness of test statistics with asymmetric distributions as in the general case when $I > 2, J > 2$. We will give a counterexample on that issue.

6.4.1 Uncorrelatedness Among Synchronized Permutation Tests

The uncorrelation among the test statistics is always a good property to have since it avoids the possibility that the activeness of one factor might influence the conclusions on the other ones. To prove the uncorrelation among all the synchronized permutation test statistics, we need to consider the implications:

(i) if $\rho(^aT_A^*, {}^bT_B^*) = 0 \Rightarrow \rho(^aT_A^*, {}^bT_{AB}^*) = 0$, $\rho(^bT_B^*, {}^aT_{AB}^*) = 0$;
(ii) if $\rho(^aT_A^*, {}^aT_{AB}^*) = 0 \Rightarrow \rho(^bT_B^*, {}^bT_{AB}^*) = 0$,

where $\rho(X, Y)$ stands for the correlation between X and Y. Let us consider the proof of (i) first. The tests on main factors are defined on different permutation spaces. In particular, $^aT_A^*$ is obtained by exchanging units between pairs of rows and within columns, and $^bT_B^*$ is obtained by exchanging units between pairs of columns and within rows. Hence, without loss of generality, consider the case where we are testing for $H_{0A} : \alpha_i = 0 \; \forall i$. This hypothesis allows us to exchange units between rows and within columns. The test statistic $^aT_A^*$ is a random variable, and its probability function is derived by exchanging units within columns, so the total of each column is constant. Let $Y_{.j}, j = 1, \ldots, J$ be the total of column j. Then

$$\sum_i T^*_{jh|i} = \sum_i \left[\sum_k y^*_{ijk} - \sum_k y^*_{ihk} \right] = \sum_i \sum_k y^*_{ijk} - \sum_i \sum_k y^*_{ihk} = Y_{.j} - Y_{.h},$$

and hence

$$^bT^*_B = \sum_{j<h} \left[\sum_i T^*_{jh|i} \right]^2 \equiv \sum_{j<h} (Y_{.j} - Y_{.h})^2 = {}^bT_B,$$

where ^bT_B is the observed value of the test statistic for factor B. Since $^bT^*_B$ is constant, $\rho(^aT^*_A, {}^bT^*_B) = 0$.

Now we will prove point (ii). Both the $^aT^*_A$ and $^aT^*_{AB}$ probability distributions are derived from the same permutations (within columns) since $^aT^*_{AB}$ is also obtained from a random synchronized permutation based on factor A. Both test statistics are based on the random variables $T^*_{is|j}$, which are differences between the sums of n independent random variables y^*_{ijk}. Since $\text{Cov}(y^*_{ijk}, y^*_{ihk}) = 0$ (i.e., two observations from two distinct columns are uncorrelated), the $T^*_{is|j}$'s are i.i.d. random variables with $\text{E}[T^*_{is|j}] = (n - 2\nu)(\alpha_i - \alpha_s + \alpha\beta_{ij} - \alpha\beta_{sj})$ and $\text{Var}[T^*_{is|j}] = 2n\sigma^2$. To begin with, consider the case where $I = J = 2$. Here we do not need the squaring operators in formulas (6.5) and (6.6) since the test statistics already depend on all the main and interaction effects. Therefore, in this case we may write

$$^aT^*_A = {}^aT^*_{12|1} + {}^aT^*_{12|2},$$
$$^aT^*_{AB} = {}^aT^*_{12|1} - {}^aT^*_{12|2}.$$

We wish to evaluate

$$\text{Cov}(^aT^*_A, {}^aT^*_{AB}) = \text{E}[^aT^*_A, {}^aT^*_{AB}] - \text{E}[^aT^*_A]\text{E}[^aT^*_{AB}].$$

Let $^aT^*_{12|1} = X$ and $^aT^*_{12|2} = Y$. Then

$$\begin{aligned}
\text{Cov}(^aT^*_A, {}^aT^*_{AB}) &= \text{E}[(X+Y)(X-Y)] - \text{E}(X+Y)\text{E}(X-Y) \\
&= \text{E}[X^2] - \text{E}[Y^2] - \text{E}(X)^2 + \text{E}(Y)^2 \\
&= \text{Var}(X) - \text{Var}(Y) = 0
\end{aligned}$$

since $\text{Var}(X) = \text{Var}(Y) = 2n\sigma^2$.

It is not easy to prove unbiasedness in the general $I \times J$ case without requiring some further assumptions on experimental errors. However, we conjecture that the uncorrelatedness among test statistics is still maintained because of the permutation structure. As a further proof, we have run some simulations of a 2^2 full factorial design with $n = 3$ runs and some different settings, including $H_{0A} \cap H_{0B}$, $H_{1A} \cap H_{0B}$, and $H_{1A} \cap H_{1B}$.

Figure 6.2 reports two scatterplots of $^aT_A^*$ vs. $^bT_B^*$ and their related p-values, respectively. We have applied the CSPs (see Section 6.3) to obtain these results. Note how the points are spread at random on the left in Figure 6.2. On the right in Figure 6.2 it is possible to see the discrete nature of the test statistic distribution when n is small and CSPs are applied. Note that the cardinality of distinct values of $^aT_A^*$ when $n = 3$ is 10. That is why the points on the right of Figure 6.2 are spread *around* the theoretical attainable points whose coordinates are $(m/10, n/10)$, with $m, n = 1, \ldots, 10$. These points are spread around the theoretical attainable points since we have applied Monte Carlo generation for the CSPs.

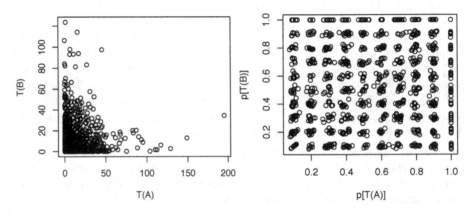

Fig. 6.2. 2×2 design with three replicates. **Left:** Scatterplot of $^aT_A^*$ vs. $^bT_B^*$. **Right:** Scatterplot of $p(^aT_A^*)$ vs. $p(^bT_B^*)$. CSPs applied, both factors under the null hypothesis.

In Figure 6.3 the same graphics are reported, but here factor A is active with $\alpha_1 = -\alpha_2 = 10\sigma$, where $\sigma = 1$. The interesting part of this figure is the scatterplot of p-values. See how the clouds of points randomly lay around the theoretical lines for which $p(^bT_B^*) = n/10$ $(n = 1, \ldots, 10)$, but they are centered on the minimum attainable p-value for factor A, which is $p(^aT_A^*) = 1/10$.

Finally, Figure 6.4 reports the same graphics when both factors are active: factor A with the same settings as before and factor B with effects set equal to 5σ in absolute value. This figure shows how the synchronized permutation test statistics for the main factors are uncorrelated even under the alternative hypothesis. On the right in Figure 6.4 it is possible to see how the points are randomly spread around $(1/10, 1/10)$, which represents the theoretical point of minimum attainable p-value for both main factors.

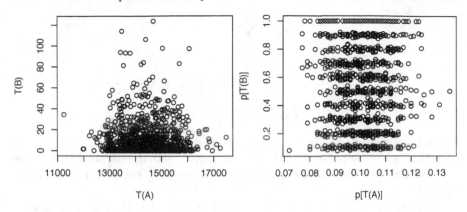

Fig. 6.3. 2×2 design with three replicates. **Left:** Scatterplot of ${}^{a}T_{A}^{*}$ vs. ${}^{b}T_{B}^{*}$. **Right:** Scatterplot of $p({}^{a}T_{A}^{*})$ vs. $p({}^{b}T_{B}^{*})$. CSPs applied, $\alpha_1 = -\alpha_2 = 10$, $\beta_1 = \beta_2 = 0$.

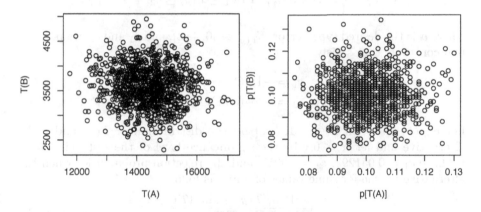

Fig. 6.4. 2×2 design with three replicates. **Left:** Scatterplot of ${}^{a}T_{A}^{*}$ vs. ${}^{b}T_{B}^{*}$. **Right:** Scatterplot of $p({}^{a}T_{A}^{*})$ vs. $p({}^{b}T_{B}^{*})$. CSPs applied, $\alpha_1 = -\alpha_2 = 10$, $\beta_1 = -\beta_2 = 5$.

6.4.2 Unbiasedness and Consistency of Synchronized Permutation Tests

A test T of size α, $\alpha \in [0,1]$, with acceptance region $A_\alpha(Y)$ and rejection region $R_\alpha(Y)$, is unbiased if $\mathrm{Pr}_{H_1}[T \in R_\alpha(Y)] \geq \alpha = \mathrm{Pr}_{H_0}[T \in R_\alpha(Y)]$. We will only give the proof of unbiasedness in the simple case when $I = J = 2$. To this end, consider the test statistics

$$^aT_A^* = \sum_{j=1}^{2} {}^aT_{12|j}^*,$$

$$^bT_B^* = \sum_{i=1}^{2} {}^bT_{12|i}^*,$$

$$^aT_{AB}^* = {}^aT_{12|1}^* - {}^aT_{12|2}^*,$$

which have symmetric distributions. The symmetry of the test statistic distribution is a core point since it is easy to prove that test statistics whose distributions are asymmetric are biased. To see this, consider the following example.

Example. Let $T(y)$ be a test statistic to test for $H_0 : \mu = 1$ against $H_1 : \mu \neq 1$, where μ is the mean of the exact distribution of $T(y)$. Moreover, suppose $T(y) \sim \exp(1)$ under H_0. Since the exact null distribution of $T(y)$ is known, it is possible to build the exact acceptance region for the test as

$$A_\alpha(T) = \left\{ T(y) \in S_{T(y)} : \Pr_{H_0}[c_1 \leq T(y) \leq c_2] = 1 - \alpha \right\},$$

where α is the desired type I error, $S_{T(y)} \equiv [0, +\infty[$, and c_1 and c_2 are positive real constants that satisfy

$$c_1 = -\log(1 - \alpha/2),$$
$$c_2 = -\log(\alpha/2).$$

Let $\exp(1 + \delta)$ with $\delta \in] - 1, +\infty[$ be the distribution of $T(y)$ under the alternative and let $\alpha = 0.1$ be the significance level of the test. Therefore, we have $c_1 = 0.051293$, $c_2 = 2.9957$, and the rejection probabilities when the alternative is true for some values of δ are as follows:

δ	$\Pr_{H_1}[T(y) \notin A_{0.1}(T)]$
0	0.1
0.1	0.0919
0.2	0.0871
0.4	0.0844
0.6	0.0870
1.	0.1

Hence T is biased since the condition $\Pr_{H_1}[T(y) \in R_\alpha(T)] \geq \Pr_{H_0}[T(y) \in R_\alpha(T)]$ is not always satisfied with $\alpha = 0.1$. This counterexample shows that it does not seem possible to prove the unbiasedness of a test whose test statistic has an asymmetric null distribution.

We will only prove the unbiasedness of $^aT_A^*$ when $I = J = 2$, as the proof for the other test statistic is similar. Suppose, without loss of generality, that $H_{0A} : \alpha_1 - \alpha_2 = 0$ vs. $H_{1A} : \alpha_1 - \alpha_2 > 0$. Hence the null hypothesis is rejected for large values of ^aT_A. Recall the permutation structure of $^aT_A^*$ and

^aT_A introduced in Section 6.2. Let us denote by $\mathbf{y}(0)$ and $\mathbf{y}(\alpha)$ the data when they are under the null hypothesis and under the alternative, respectively. If H_{1A} is true, then

$$^aT_A^* = 2(n - 2\nu)[\alpha_1 - \alpha_2] + \sum_{j=1}^{2} \bar{\varepsilon}_{12|j}^*,$$

$$^aT_A = 2n[\alpha_1 - \alpha_2] + \sum_{j=1}^{2} \bar{\varepsilon}_{12|j}.$$

Therefore $\Pr\left[^aT_A^* \geq {}^aT_A | \mathbf{y}(\alpha)\right]$ is equal to

$$\Pr\left[2(n - 2\nu)[\alpha_1 - \alpha_2] + \sum_{j=1}^{2} \bar{\varepsilon}_{12|j}^* \geq 2n[\alpha_1 - \alpha_2] + \sum_{j=1}^{2} \bar{\varepsilon}_{12|j}\right]$$

$$= \Pr\left[\sum_{j=1}^{2} \bar{\varepsilon}_{12|j}^* \geq \sum_{j=1}^{2} \bar{\varepsilon}_{12|j} + 4\nu(\alpha_1 - \alpha_2)\right].$$

If H_{0A} is true, then

$$\Pr\left[^aT_A^* \geq {}^aT_A | \mathbf{y}(0)\right] = \Pr\left[\sum_{j=1}^{2} \bar{\varepsilon}_{12|j}^* \geq \sum_{j=1}^{2} \bar{\varepsilon}_{12|j}\right].$$

Therefore

$$\Pr\left[^aT_A^* \geq {}^aT_A | \mathbf{y}(\alpha)\right] \leq \Pr\left[^aT_A^* \geq {}^aT_A | \mathbf{y}(0)\right],$$

and hence the p-values when H_{1A} is true are not greater than the p-values when H_{0A} is true since $\Pr[\nu = 0] < 1$. This proves the unbiasedness of $^aT_A^*$. Of course, a similar result can be obtained for unrestricted alternatives.

The same arguments can be used to prove the consistency of $^aT_A^*$. By definition, a test is consistent if $\lim_{n\to+\infty} \Pr_{H_1}[T(y) \in R_\alpha(T)] = 1$. This means that, as the available information increases ($n \to \infty$), the test can distinguish with probability one even small departures from the null hypothesis. To prove consistency, first note that the intermediate statistic for testing factor A, $\sum_j T_{is|j}^*$, is consistent for restricted alternatives when $I = 2$ and $J > 2$. The proof is similar to that of unbiasedness since

$$\Pr\left[\sum_j T_{is|j}^* \geq \sum_j T_{is|j} | \mathbf{y}(\alpha)\right] = \Pr\left[\sum_j \bar{\varepsilon}_{is|j}^* \geq \sum_j \bar{\varepsilon}_{is|j} + 2J\nu(\alpha_i - \alpha_s)\right]$$

$$= \Pr\left[\sum_j \frac{\bar{\varepsilon}_{is|j}^*}{2nJ} \geq \sum_j \frac{\bar{\varepsilon}_{is|j}}{2nJ} + \frac{\nu}{n}(\alpha_i - \alpha_s)\right].$$

Now let $n \to +\infty$. Recall the model assumptions $E[\varepsilon_{ijk}] = 0$, $V[\varepsilon_{ijk}] = \sigma^2 < +\infty$. The random variable $(2nJ)^{-1} \sum_j \bar{\varepsilon}^*_{is|j}$ satisfies

$$E\left[\frac{1}{2nJ} \sum_j \bar{\varepsilon}^*_{is|j}\right] = 0,$$

$$\lim_{n \to +\infty} \text{Var}\left[\frac{1}{2nJ} \sum_j \bar{\varepsilon}^*_{is|j}\right] = 0,$$

whereas, if H_{1A} is true, we have

$$\lim_{n \to +\infty} \text{Pr}\left[\frac{\nu}{n}(\alpha_i - \alpha_s) > 0\right] = 1$$

because $\alpha_1 - \alpha_2 > 0$, $0 \le \nu \le n$, and $\lim_{n \to +\infty} \text{Pr}_\nu[\nu = 0] = 0$ by applying either CSPs or USPs (see Section 6.3). Therefore,

$$\lim_{n \to +\infty} \text{Pr}\left[\sum_j T^*_{is|j} \ge \sum_j T_{is|j} | \mathbf{y}(\alpha)\right] = 0.$$

Hence, as n increases, the p-value tends to zero in probability, so the probability of rejecting the null hypothesis tends to one. To complete the proof, note that $\left[\sum_j T^*_{is|j}\right]^2$ is consistent for unrestricted alternatives, and the sum of consistent tests is also consistent. Therefore

$$^a T^*_A = \sum_{i<s} \left[\sum_j T^*_{is|j}\right]^2$$

is consistent.

6.5 Power Simulation Study

In this section, we report a simulation study performed with the goal of validating the synchronized permutation testing on $I \times J$ replicated factorial designs. All the examples regard a 3×2 design, and the emphasis is on several aspects of the proposed procedure: (i) the behavior of CSP/USP procedures under H_0 and in power for the two-level factor and the three-level factor; (ii) the behavior of CSP/USP testing procedures for the main factors and interaction in a power simulation with different kinds of error distributions; (iii) the fact that the statistics for main factors and interaction are uncorrelated. For each aspect, a comparison with the parametric ANOVA counterpart is also reported. Figure 6.5 reports the behavior of synchronized permutation testing and ANOVA on a 3×2 design with $n = 3, 5, 10$, with some different settings, H_{0A}, H_{0B}, H_{1A}, and H_{1B}, when the errors are normally distributed. The interaction has not been considered in order to focus the attention on the main factor tests. However, we note that the proposed test statistics are uncorrelated, and hence the presence of interaction is irrelevant to our aim. In the power simulation, the effects of each main factor were set equal to $\sigma/2$ (where $\sigma^2 = 1$ is the error variance), in accordance with formula (6.1). The graphs have been obtained as follows. For each setting, we generated 1000 independent samples, applied the testing procedure, and stored the p-values of the main factors. Then, for each main factor and each setting, we represented the cumulative distribution functions of the related p-values. We see that there is a discrepancy in the c.d.f.s of the p-values between the two-level factor and the three-level factor when the null hypotheses H_{0A} and H_{0B} are true (solid lines). In particular, when $n = 3$, the p-value c.d.f. of the two-level factor is very close to the hypothetical continuous uniform distribution, while this is not true for the three-level factor. Obviously, this gap affects the related performances in power (dotted lines). This point needs some further investigation. This behavior might be due to the fact that USPs do not account for dependencies among partial test statistics (cfr. Subsection 5.2.1). That is why we did not consider the USP testing procedure in the power comparison (see Table 6.1). However, the gap between the two c.d.f.s tends to vanish as n increases. For $n = 5$, the exact CSP procedure has been applied, and in all the remaining cases we considered 1000 Monte Carlo synchronized permutations. The bottom of Figure 6.5 represents the comparison with the ANOVA statistic. Note that here we have the same behavior in power with respect to permutation tests for factor A and factor B, respectively (dotted lines), though the error rates under the relative null hypotheses (solid lines) do not show any significant departure from the hypothetical continuous uniform c.d.f.

Table 6.1 reports a comprehensive power simulation comparison between the CPS testing procedure and the two-way ANOVA test. Four types of error distributions have been considered: the normal distribution for continuity with the literature, the Gamma (with one and two degrees of freedom) representing asymmetric error distribution, and the Student t_3 distribution representing heavy-tailed errors. The first column reports the number of replicates considered in each setting. The nominal significance levels in this study have been chosen to be as close as possible to the usual ones (1%, **5%**, and *10%*) from the achievable significance levels available when applying the CSPs: When $n = 4$, the nominal levels are 0.029, **0.058**, and *0.114*; when $n = 7, 10$, the nominal levels are 0.016, **0.048**, and *0.104*. Recall that the achievable significance levels are multiples of $2/C_{CSP}$. Then, for each error distribution, the observed rejection rates are reported for the main factors and interaction at the corresponding nominal level. The effects of main factors and interaction were set in accordance with a 3×2 full factorial design. The true sizes of the effects (whose label is "T.S.") are displayed in the third column and were

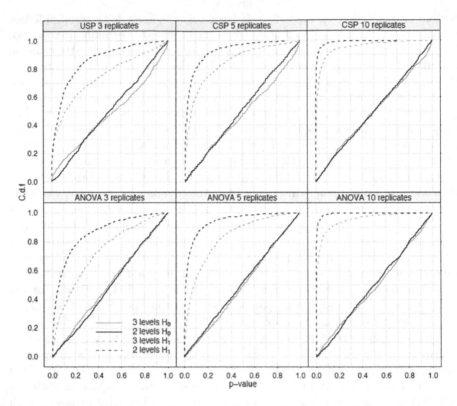

Fig. 6.5. Simulation study for the main effects of synchronized permutation tests and ANOVA in a 3×2 design under H_0 and under H_1. Continuous lines represent the type I error rate and dashed lines the statistical power of the test.

set equal to $\sigma/2, 3\sigma/2, \sigma$ for factor A, factor B, and interaction, respectively. The error variance σ^2 is held fixed to one. As regards interaction, we have considered the test statistic T_{AB}^{**} illustrated at the end of Section 6.2. Note how the power of synchronized permutation tests is very close to that of the parametric ANOVA test for any distribution considered.

Finally, in Figure 6.6, we empirically show that the behavior of the test statistic for one main factor is not affected by the activeness of the other factor. In the figure, we plotted the observed power of the test on factor A as the size of its effects increases from 0 to 1.5σ in steps of $.25\sigma$ in a 3×2 full factorial design. Each line in the graph represents the observed power for factor A at a significance level of 5% at different sizes of the factor B effects. The data in the simulation include an interaction effect of size σ, and each point is obtained by 1000 independent data generations. As in Figure 6.5, the simulation was run with $n = 3, 5, 10$, and a comparison with the ANOVA is shown in the bottom of the same figure. Note that, regardless of the size

Table 6.1. Power simulation in a 3×2 design. Synchronized permutation and two-way ANOVA tests.

n	Factor	P_ε T.S.	Norm			Exp			t_3			Ga_2		
			Constrained Synchronized Permutation Tests											
	A	.5	.226	.353	.507	.309	.442	.579	.360	.516	.664	.278	.405	.548
4	B	1.5	.996	.999	1.00	.996	.998	1.00	.983	.992	.996	.992	.998	1.00
	AB	1	.814	.906	.958	.857	.906	.960	.867	.907	.953	.850	.922	.964
	A	.5	.387	.606	.763	.439	.615	.734	.514	.685	.797	.426	.620	.741
7	B	1.5	1.00	1.00	1.00	.615	1.00	1.00	.997	.999	1.00	1.00	1.00	1.00
	AB	1	.997	.998	1.00	.981	.992	.998	.974	.981	.994	.988	.997	1.00
	A	.5	.521	.691	.796	.523	.678	.794	.592	.746	.841	.482	.652	.792
10	B	1.5	1.00	1.00	1.00	1.00	1.00	1.00	.999	1.00	1.00	1.00	1.00	1.00
	AB	1	1.00	1.00	1.00	.997	.998	1.00	.992	.993	.994	.999	1.00	1.00
n	Factor	T.S.	Two-Way ANOVA Tests											
	A	.5	.244	.373	.516	.344	.455	.582	.401	.524	.650	.299	.409	.544
4	B	1.5	1.00	1.00	1.00	1.00	1.00	1.00	.994	.997	.997	1.00	1.00	1.00
	AB	1	.871	.928	.961	.845	.914	.949	.882	.920	.951	.856	.919	.959
	A	.5	.431	.636	.763	.427	.613	.736	.517	.686	.783	.433	.614	.755
7	B	1.5	1.00	1.00	1.00	1.00	1.00	1.00	.958	.999	1.00	1.00	1.00	1.00
	AB	1	.955	.997	1.00	.965	.992	.995	.974	.987	.993	.981	.994	1.00
	A	.5	.658	.803	.892	.663	.795	.877	.749	.847	.907	.637	.796	.880
10	B	1.5	1.00	1.00	1.00	1.00	1.00	1.00	1.00	1.00	1.00	1.00	1.00	1.00
	AB	1	1.00	1.00	1.00	.995	.997	1.00	.987	.992	.993	.999	1.00	1.00

of the factor B effect, the corresponding lines are very close to each other, showing that no influence on the performance of the factor A test statistic is due to factor B effect sizes. This is also true for interaction (although we do not report a similar graph here) since the corresponding test statistics ($^aT^*_{AB}$ or $^bT^*_{AB}$) only depend on interaction effects.

6.6 Multiple Comparisons

Once the presence of main effects has been determined, a researcher may be interested in the *post-hoc* analysis; that is, in examining the statistical incidence of each single effect. In what follows, we will always refer to the effects of the row factor (factor A). In an *all-pairwise comparison*, each level of a factor is compared with every other level and a family of $C = I(I-1)/2$ null hypotheses are tested. The null hypotheses, for example, concern the differences of pairs of true effects α_i and α_s with $i \neq s$; that is, $H_{0A}^{is} : \{\alpha_i = \alpha_s\}$ for all $1 \leq i < s \leq I$. This entails testing a family of $C = I(I-1)/2$ minimal dependent hypotheses.

Multiple comparisons can be carried out by computing *simultaneous confidence intervals* that make it possibile to represent the results graphically

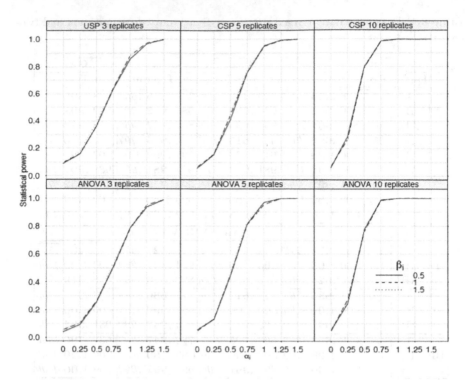

Fig. 6.6. Statistical power at a significance level of 5% for different intensities of the effects α_i and β_i.

and thus infer the practical and statistical significance of differences by visual inspection (Hsu, 1996). Confidence intervals are determined for the differences between the observed row means $\bar{y}_{i..} = \sum_{jk} y_{ijk}/nJ$ and $\bar{y}_{s..} = \sum_{jk} y_{sjk}/nJ$ of any pair of levels i and s with mean response α_i and α_s. Then, the differences are declared significant if $|\bar{y}_{i..} - \bar{y}_{s..}| > \text{MSD}$, where MSD is the *minimum significant difference* derived from the corresponding confidence interval. Well-known parametric methods for computing MSDs are, for example, *Scheffé's method* or the more powerful *Tukey's honest significant difference method* (Hsu, 1996; Dean and Voss, 1999). An algorithm for the permutation approach for computing MSDs between two treatments i and s is given in Algorithm 1. To the best of our knowledge, no attempt has been made to derive simultaneous confidence intervals for multiple comparisons from permutation tests. In the following, we propose two procedures based on the algorithm given by Pesarin (2001) for computing the confidence interval of the difference between two treatments. Algorithm 1 computes the half confidence interval for the difference between two effects of one factor (hence corresponding to the MSD) by increasing an initial MSD value provided by one of the parametric methods such as, for example, from the t test $\text{MSD} = t_{\alpha/C, I(n-1)} \cdot \sqrt{2 \cdot \text{MSE}/nJ}$ (where

MSE is the mean square error), and in that case, made slightly smaller. The same procedure could then be repeated for the other half interval, although here we confine ourselves to considering symmetric intervals.

Procedure Compute_MSD(i, s);

1. Choose a negative MSD_{is} and the desired precision ϵ.
2. Subtract $\bar{y}_{i\cdot\cdot} - \bar{y}_{s\cdot\cdot} + \text{MSD}_{is}$ from each value of the data group related to one of the two factor levels considered, say the ith, obtaining the new set of observations
$$\tilde{y}_{ijk} = y_{ijk} - \bar{y}_{i\cdot\cdot} + \bar{y}_{s\cdot\cdot} - \text{MSD}_{is}; \ j = 1, ..., J; \ k = 1, ..., n.$$
These observations will then be pooled with the original observations $\tilde{y}_{sjk} = y_{sjk}$ to apply the synchronized permutations.
3. Compute the observed test statistic for the new responses:
$$T^{obs} = \tilde{\bar{y}}_{i\cdot\cdot} - \tilde{\bar{y}}_{s\cdot\cdot} = -\text{MSD}_{is}.$$
4. By rearranging the observations B times with *synchronized permutations* (CSPs or USPs) within the J columns, obtain the permutation distribution of the statistic
$$T^*(\text{MSD}_{is}) = \tilde{\bar{y}}^*_{i\cdot\cdot} - \tilde{\bar{y}}^*_{s\cdot\cdot}.$$
5. If the condition
$$|\#\{T^*(\text{MSD}_{is}) \le T^{obs}\}/B - \alpha/2| < \epsilon/2$$
is satisfied, $|\text{MSD}_{is}|$ is the desired halfwidth of the $(1 - \alpha)\%$ confidence interval for $\alpha_i - \alpha_s$. Otherwise, increase $|\text{MSD}_{is}|$ and repeat steps $2 \to 4$ until condition 5 is satisfied.

Algorithm 1: The algorithm to compute the halfwidth of the confidence interval of the difference between two effects.

There are two possible ways to use Algorithm 1 in the computation of simultaneous confidence intervals for C comparisons. The first procedure seeks the width of the interval that satisfies all comparisons simultaneously at a confidence level of $1 - \alpha$. We will call the confidence intervals thus obtained "simultaneous". They can be obtained as follows. Let MSD_{is} be the value found by Algorithm 1 for treatments i and s, and consider MSD_{ir}, the next comparison between treatments i and r. Initially, we set $\text{MSD}_{ir} = \text{MSD}_{is}$. Then, a procedure similar to Algorithm 1 is used to determine whether MSD_{ir} guarantees the confidence level $1 - \alpha$ or not. In the former case, we may stop and proceed to the next comparison (note that MSD_{ir} may also guarantee more than the confidence $1 - \alpha$), while in the latter case we increase MSD_{ir} until the confidence level is guaranteed. In this way, the final half confidence interval for all comparisons becomes

$$\text{MSD} = \sup_{i<s} \text{MSD}_{is};$$

that is, MSD is half of the simultaneous confidence interval for the difference of two effects derived by the synchronized permutation test that guarantees a coverage of $1 - \alpha$. A graphical representation of simultaneous confidence intervals derived with this procedure is given on the left in Figure 6.7. The graph is obtained by attaching error bars to a scatterplot of the estimated effects versus treatment labels. The lengths of the error bars are adjusted so that the population means of a pair of treatments can be inferred to be different if their bars do not overlap. From the definition of MSD, the bars correspond to $\bar{y}_{i\cdot\cdot} \pm \text{MSD}/2$. In fact, given $\bar{y}_{i\cdot\cdot} < \bar{y}_{j\cdot\cdot}$, then $\bar{y}_{j\cdot\cdot} - \text{MSD}/2 < \bar{y}_{i\cdot\cdot} + \text{MSD}/2$ iff $|\bar{y}_{j\cdot\cdot} - \bar{y}_{i\cdot\cdot}| < \text{MSD}$. That is, the bars of the confidence intervals relative to two effects, α_i and α_j, overlap iff the difference between the two corresponding means is smaller than the MSD.

The second procedure consists of computing different confidence intervals for each comparison by using Algorithm 1 with confidence level adjusted in order to control the error for the multiple testing. The confidence intervals obtained in this way will be called "individual", as the computation of the confidence interval for a single comparison does not take into account the role of the other comparisons. A basic method treats all comparisons as independent and consequently adjusts the error rate per comparison as $\alpha_{PC} = 1 - (1 - \alpha)^{1/C}$. More simply, given that $(1 - \alpha)^{1/C} < 1 - \alpha/C$ and the difference is small, $\alpha_{PC} = \alpha/C$ may also be used. This latter is commonly known as the *Bonferroni adjustment* . In this case, a different graphical representation is required, as the previous one works only for tests where the simultaneous confidence intervals have the same widths for all individual comparisons. An alternative representation is suggested by Hsu (1996) and is shown on the right in Figure 6.7. It consists of a two-dimensional space in which a 45° line represents the points satisfying $\bar{y}_{i\cdot\cdot} = \bar{y}_{s\cdot\cdot\cdot}$. At each point $(\bar{y}_{i\cdot\cdot}, \bar{y}_{s\cdot\cdot})$, for which coordinates are given by the sample means of two levels, a segment of slope -1 is drawn, centered in $(\bar{y}_{i\cdot\cdot}, \bar{y}_{s\cdot\cdot})$ and of length $\text{MSD}/\sqrt{2}$. Statistical inference is derived by checking whether the line segment crosses the 45° line. The practical assessment of mean differences is preserved instead on the x-axis or y-axis. All the C confidence intervals can be represented by drawing only the segments with $\bar{y}_{i\cdot\cdot} > \bar{y}_{s\cdot\cdot}$ (i.e., only intervals below the 45° line).

We performed a simulation study for validating these two new procedures for simultaneous confidence intervals against the parametric Tukey's honest significant method. The goal of the study is also the empirical comparison of the two procedures (individual and simultaneous). Since the aim of confidence interval representation is to have direct information about which comparisons are significant, we generated data under H_0. Similarly to the study of the previous section, we considered four different distributions of errors. To check whether the intervals thus constructed contain the true value of the mean differences or not, we generate 1000 new samples under H_0 per distribution. For each distribution, we compute the halfwidth MSD_{is} of the confidence interval with the two procedures described at each generation of data. Hence, for the simultaneous confidence intervals, the MSD_{is} are all equal, while for the individual confidence intervals, they vary for each pair of treatments.

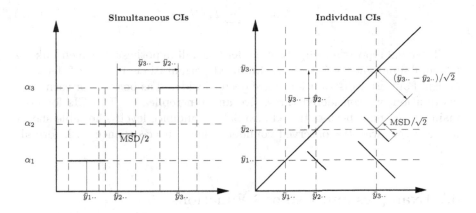

Fig. 6.7. Simultaneous (left) and individual (right) confidence interval graphical representations.

For each generation $l = 1, \ldots, 1000$, and each comparison, we define the confidence intervals (CIs) as:

$$\mathrm{CI}_{is} : \bar{y}_{i\cdot\cdot} - \bar{y}_{s\cdot\cdot} \pm \mathrm{MSD}_{is}, \qquad 1, \leq i < s \leq I,$$

and set $P_l = 1$ if all the CIs contain the zero and $P_l = 0$ otherwise.

Table 6.2. Confidence levels achieved in multiple comparisons.

3 Replicates	Norm		Exp		t3		Ga₂	
	A	B	A	B	A	B	A	B
Tukey's CIs	0.983	0.980	0.555	0.678	0.969	0.961	0.975	0.975
Simultaneous CIs	0.984	0.999	0.673	0.754	0.975	0.987	0.974	0.975
Individual CIs	0.989	0.998	0.431	0.754	0.978	0.987	0.761	0.975
5 Replicates	**Norm**		**Exp**		**t3**		**Ga₂**	
	A	B	A	B	A	B	A	B
Tukey's CIs	0.953	0.951	0.940	0.943	0.893	0.907	0.961	0.965
Simultaneous CIs	0.962	0.974	1.000	0.926	0.873	0.976	0.974	0.976
Individual CIs	0.916	0.974	0.938	0.926	0.688	0.976	0.792	0.976
10 Replicates	**Norm**		**Exp**		**t3**		**Ga₂**	
	A	B	A	B	A	B	A	B
Tukey's CIs	0.961	0.960	0.884	0.911	0.975	0.977	0.767	0.789
Simultaneous CIs	0.946	0.992	0.844	0.837	0.954	0.984	0.779	0.746
Individual CIs	0.968	0.991	0.806	0.812	0.971	0.983	0.749	0.751

Finally, the confidence achieved is given by:

$$\text{Conf} = \sum_{l=1}^{1000} P_l/B. \qquad (6.13)$$

Table 6.2 reports a comparison under the null hypothesis between Tukey's CIs and simultaneous CIs of synchronized permutations in a 3×2 factorial design. Four error distributions have been considered as before. For each error distribution, we considered three, five and ten replicates. The CIs for both main factors have been obtained at a 95% confidence level. The table shows, for each scenario, the observed confidence level in accordance with formula (6.13).

6.7 Examples and Use of R Functions

This section is devoted to applying the synchronized permutation tests to some examples from Montgomery (2001) and to showing how to apply the R functions that have been implemented to perform the synchronized permutation tests. The R examples and the results from Montgomery (2001) have been obtained by computing the test statistics introduced in Section 6.2 and applying a Monte Carlo algorithm to randomly sample from the CSP/USP distribution in accordance with Section 6.3. For each example, we also report the corresponding two-way ANOVA results. In some cases, the exact CSP distribution has also been considered.

To begin with, let us introduce a simple generic example in order to better understand how synchronized permutations really work. Suppose that we are dealing with a 2^2 complete factorial design with $n = 2$. This can be displayed in table format:

	Factor B	
Factor A	$y_{111}\ y_{112}$	$y_{121}\ y_{122}$
	$y_{211}\ y_{212}$	$y_{221}\ y_{222}$

Suppose we are testing for factor A and wish to apply CSPs in order to obtain the synchronized permutation distribution of the statistic. Recall that when CSPs are applied, units are swapped in the same original position of the first couple of blocks. Therefore, we will focus on the swaps from blocks A_1B_1 and A_2B_1. Here there are $\binom{4}{2} = 6$ distinct permutations:

(1) no swaps (i.e. no permutation);
(2) swap y_{111} with y_{211};
(3) swap y_{111} with y_{212};
(4) swap y_{112} with y_{211};
(5) swap y_{112} with y_{212};

(6) swap both y_{111} and y_{112} with y_{211} and y_{212}.

Clearly, permutations (1) and (6) give the same value of the statistic (i.e., $^aT_A[1] = {}^aT_A$). Permutations (2) and (5) lead to the same value of the statistic (say $^aT_A[2]$), as well as permutations (3) and (4) (say $^aT_A[3]$). Therefore, according to this example, the minimum attainable significance level is $1/3$. Instead, if USPs are adopted, there are $\sum_{\nu=0}^{2} \binom{2}{\nu}^4 = 18$ distinct synchronized permutations, which are listed in Table 6.3 (in the π index column).

Table 6.3. List of all possible USPs in a 2^2 design with $n = 2$.

π index	ν	1st colum	2nd colum	$^aT_A^*$
1	0	$\emptyset$	$\emptyset$	$^aT_A^*[1]$
2		$y_{111} \longleftrightarrow y_{211}$	$y_{121} \longleftrightarrow y_{221}$	$^aT_A^*[2]$
3		$y_{111} \longleftrightarrow y_{211}$	$y_{121} \longleftrightarrow y_{222}$	$^aT_A^*[3]$
4		$y_{111} \longleftrightarrow y_{211}$	$y_{122} \longleftrightarrow y_{221}$	$^aT_A^*[4]$
5		$y_{111} \longleftrightarrow y_{211}$	$y_{122} \longleftrightarrow y_{222}$	$^aT_A^*[5]$
6		$y_{111} \longleftrightarrow y_{212}$	$y_{121} \longleftrightarrow y_{221}$	$^aT_A^*[6]$
7		$y_{111} \longleftrightarrow y_{212}$	$y_{121} \longleftrightarrow y_{222}$	$^aT_A^*[7]$
8		$y_{111} \longleftrightarrow y_{212}$	$y_{122} \longleftrightarrow y_{221}$	$^aT_A^*[8]$
9		$y_{111} \longleftrightarrow y_{212}$	$y_{122} \longleftrightarrow y_{222}$	$^aT_A^*[9]$
	1			
10		$y_{112} \longleftrightarrow y_{211}$	$y_{121} \longleftrightarrow y_{221}$	$^aT_A^*[9]$
11		$y_{112} \longleftrightarrow y_{211}$	$y_{121} \longleftrightarrow y_{222}$	$^aT_A^*[8]$
12		$y_{112} \longleftrightarrow y_{211}$	$y_{122} \longleftrightarrow y_{221}$	$^aT_A^*[7]$
13		$y_{112} \longleftrightarrow y_{211}$	$y_{122} \longleftrightarrow y_{222}$	$^aT_A^*[6]$
14		$y_{112} \longleftrightarrow y_{212}$	$y_{121} \longleftrightarrow y_{221}$	$^aT_A^*[5]$
15		$y_{112} \longleftrightarrow y_{212}$	$y_{121} \longleftrightarrow y_{222}$	$^aT_A^*[4]$
16		$y_{112} \longleftrightarrow y_{212}$	$y_{122} \longleftrightarrow y_{221}$	$^aT_A^*[3]$
17		$y_{112} \longleftrightarrow y_{212}$	$y_{122} \longleftrightarrow y_{222}$	$^aT_A^*[2]$
18	2	$\mathbf{y'_{11} \Longleftrightarrow y'_{21}}$	$\mathbf{y'_{12} \Longleftrightarrow y'_{22}}$	$^aT_A^*[1]$

Here it is easier to see the symmetry induced by the squaring operator in the test statistic (6.5). Clearly, permutation 1 (exchanging $\nu = 0$ units between two blocks, symbol "$\emptyset$") is equivalent to permutation 18 (exchanging all the units, $\nu = 2$, symbol "$\Longleftrightarrow$") since these permutations produce the observed value of the test statistic $^aT_A^*(1) = {}^aT_A$. The remaining unconstrained synchronized permutations ($2 \to 17$), where $\nu = 1$ (symbol "$\longleftrightarrow$"), give eight distinct values of the distribution of $^aT_A^*$. This is the starting point to calculate the cardinality of the support of $^aT_A^*$ for generic I, J, and n. The cardinality of the support of the test statistics depends on how the synchronization is done: Note that to have a synchronized permutation it is necessary that the same number of units be exchanged in each pair of rows or columns considered. It is also possible to synchronize permutations with respect to all possible pairs of rows. That is how formulas (6.9) and (6.12) have been obtained. It is worth noting that the USP test maintains its properties even if ν is allowed to change independently in each single pair of rows or columns. In this case the cardinality of the support of the test statistic would change accordingly. For instance, when $I > 2$, $J = 2$, and the synchronization is applied independently to each pair of rows, the cardinality of the support of $^aT_A^*$ becomes respectively

$$C_{USP}^o = \left[\sum_{\nu=0}^{n/2-1} \binom{n}{\nu}^{2J} + \frac{1}{2}\binom{n}{n/2}^{2J} \right]^{I(I-1)/2},$$

$$C_{USP}^e = \left[\sum_{\nu=0}^{(n-1)/2} \binom{n}{\nu}^{2J} \right]^{I(I-1)/2},$$

since with this design there are $I(I-1)/2$ ways to have a 2×2 ANOVA table. The cardinality of the support of the USP test statistic depends on:

- the design settings (I, J, and n),
- which factor is considered (row or column synchronized permutations),
- at which level the synchronization is done (exchange ν units between all blocks in all possible pairs of rows or columns, or exchange ν units independently in each pair of rows or columns),

and it can be obtained by decomposing the $I \times J$ ANOVA design into multiple $I \times 2$ or $2 \times J$ ANOVA designs.

6.7.1 Applications with R Functions

This paragraph is dedicated to applying some R functions to perform the synchronized permutation testing. There are two functions performing the CSP and USP tests, which are respectively called CSP.r and USP.r. Another function, synchro_summary.r, gives the summary of the previous tests. Finally, the functions IC_USP.r and IC_CSP.r give the $100(1-\alpha)\%$ synchronized permutation confidence interval for a pair of row or column levels, and the function hsu.r gives the representation of the (individual) CIs above.

In order to use these functions, first put them all in the same folder (for instance, C:/Synchro), and set the working directory by typing

```
> setwd("C:/Synchro")
```

Then, to load a function (e.g., the IC_USP.r function), type

```
> source("IC_USP.r")
```

Note that the names of the function files (to be loaded in the R environment) are the same as the names of the functions to be run except, in some cases, an underscore "_" is replaced by a point (i.e., the name of the function file is IC_USP.r, but the function name is IC.USP).

Let's begin with `CSP.r` and `USP.r`. These functions require as entries a vector of length nIJ with the observed data y and an $(nIJ) \times 2$ matrix of labels x. The latter matrix is the design matrix, which specifies the levels of factor A (first column of x) and factor B (second column). For instance, if we have to test a 3×2 design with $n = 2$ replicates and we wish to apply USPs to some simulated data (e.g., normally distributed), we may proceed as follows:

```
> set.seed(1000)
> I <-3
> J<-2
> n<-2
> y<-rnorm((n*I*J))
> x1<-factor(rep(seq(1,I),each=(n*J)))
> x2<-factor(rep(rep(c(1,2),each=n),I))
> x<-data.frame(A=x1,B=x2)
> x
    A B
1   1 1
2   1 1
3   1 2
4   1 2
5   2 1
6   2 1
7   2 2
8   2 2
9   3 1
10  3 1
11  3 2
12  3 2
> round(y,digits=5)
 [1] -0.44578 -1.20586  0.04113  0.63939 -0.78655 -0.38549
 [7] -0.47587  0.71975 -0.01851 -1.37312 -0.98243 -0.55449
```

Note that the data in y are under the global null hypothesis (i.e., $\alpha_i = 0 \; \forall i$, $\beta_j = 0 \; \forall j$, $\alpha\beta_j = 0 \; \forall i, j$). There is no special requirement on the labels, we could have used -1, 0, 1 instead of 1, 2, 3. To run the USP function, type

```
> t<-USP(y,x)
```

The object t now contains the values of the `USP.r` function. To see the list of returned values, type

```
> str(t)
List of 13
 $ pa    : num 0.512
 $ pb    : num 0.11
 $ pab   : num 0.465
 $ pab.a : num 0.443
```

```
$ pab.b   : num 0.528
$ TA      : num 7.83
$ TB      : num 13.0
$ TAB.a   : num 9.41
$ TAB.b   : num 9.41
$ type    : chr "Unconstrained"
$ C       : num 1000
$ min.sig: num [1:2] 0.00195 0.03030
$ exact   : logi FALSE
```

In detail, TA, TB, TAB.a, and TAB.b are the observed values of the test statistics (6.5), (6.7), (6.6), and (6.8); pa, pb, pab.a, and pab.b are their related p-values. pab is the global p-value of the direct combining function $T_{AB} = {}^aT_{AB} + {}^bT_{AB}$. The USP.r function also returns the type of synchronization that has been applied ("Unconstrained" for USP.r and "Constrained" for CSP.r) and the number of Monte Carlo simulations, which are set equal to 1000 by default. Finally, min.sig is the minimum (theoretical) achievable significance level for the row factor (A) and the column factor (B). The value exact specifies if the permutation distribution is exact (CSP.r only) or not.

To see the results, run the function synchro_summary.r, which requires as entry an object produced either by USP.r (as the object t) or CSP.r. To summarize the results, type

```
> synchro.summary(t)
```

```
Monte Carlo Unconstrained Synchronized Permutation Testing
for Two-way ANOVA
```

Source	T	Pr(>|T|)
Factor A	7.834854	0.512
Factor B	12.98004	0.11
Interaction (a)	9.414594	0.443
Interaction (b)	9.414594	0.528
Interaction AB	18.82919	0.465

```
Signif. codes:0 '***' 0.001 '**' 0.01'*' 0.05 '.' 0.1 ' ' 1

Number of random permutations considered: 1000
Minimum achievable significance levels:
(a): 0.001949318        (b): 0.03030303
```

This output is similar to the summary() function for linear models in R. Note that the first line of output says that we have applied Monte Carlo USPs (since no algorithm to obtain exact USPs has been provided, whereas the CSP.r function can provide the exact null distribution of the test statistic). The number of random permutations (set equal to 1000 by default) can be specified as an entry of the `USP.r` function by typing "`USP(y,x,C=n.perms)`", where `n.perms` is the desired number of permutations. Finally, the line with `Minimum achievable significance levels` reports the theoretical minimum achievable significance levels, which are equal to $1/C_{USP}^{e/o}$. In order to make a comparison with the parametric F test for two-way ANOVA (here the assumptions of this test are all satisfied), type

```
> attach(x)
> fit<-aov(y~A*B)
> summary(fit)
            Df  Sum Sq Mean Sq F value Pr(>F)
A            2 0.65290 0.32645  0.8621 0.4687
B            1 1.08167 1.08167  2.8565 0.1420
A:B          2 0.78455 0.39227  1.0359 0.4107
Residuals    6 2.27205 0.37867
```

It does not make sense to apply a CSP test to the previous example because the cardinality of the support of the test statistics is too small (there are only three distinct values when $n = 2$). Let's consider another example with simulated data. We are going to simulate data from a 3×2 design with $n = 4$ and with factor A and interaction effects under the alternative. Suppose the true effects of factor A and interaction are as in Table 6.4.

Table 6.4. :True effects for the simulation in power.

$\alpha\beta$	$\beta_1 = 0$	$\beta_2 = 0$
$\alpha_1 = -1$	-0.2	+0.2
$\alpha_2 = 0$	0	0
$\alpha_3 = +1$	+0.2	-0.2

We will use some of the previous objects (I and J) to define the new simulation. A number of replicates equal to 4 is enough to apply the exact CSP testing procedure:

```
> set.seed(10)
> source("CSP.r")
> n<-4
> x1<-factor(rep(seq(1,I),each=(n*J)))
> x2<-factor(rep(rep(c(1,2),each=n),I))
> x<-data.frame(A=x1,B=x2)
> y<-rnorm(n*I*J)
> alpha=rep(c(-1,0,1),each=2*n)
> alpha.beta<-c(rep(c(-1,1),each=n),rep(0,each=(2*n)),
+ rep(c(1,-1),each=n))/5
```

```
> y<-y+alpha+alpha.beta
> round(y,digits=5)

 [1] -1.18125 -1.38425 -2.57133 -1.79917 -0.50545 -0.41021
 [7] -2.00808 -1.16368 -1.62667 -0.25648  1.10178  0.75578
[13] -0.23823  0.98744  0.74139  0.08935  0.24506  1.00485
[19]  2.12552  1.68298  0.20369 -1.38529  0.12513 -1.31906
> t<-CSP(y,x,exact=TRUE)

loading the required package: combinat

> synchro.summary(t)

Exact Constrained Synchronized Permutation Testing
for Two-Way ANOVA


Source                  T               Pr(>|T|)

Factor A                347.3366        0.028571        *
Factor B                8.879206        0.457143


Interaction (a)         188.9874        0.057143        .
Interaction (b)         188.9874        0.057143        .
Interaction AB          377.9749        0.028571        *


Signif. codes:0 '***' 0.001 '**' 0.01 '*' 0.05 '.' 0.1' ' 1


Cardinality of S(T): 35
Minimum achievable significance levels:
(a): 0.02857143        (b): 0.02857143
```

Here the minimum achievable significance level is equal to $1/35$ for both row and column synchronized permutations.

The output produced by **synchro_summary.r** is as before, with some exceptions regarding the type of test that has been applied. Note that there is the voice "Cardinality of S(T)" (which stands for "cardinality of the support of the test statistic") instead of "Number of random permutations considered", and that the p-value of factor A is the minimum achievable p-value. Note also that the combined test for interaction is more significant than the partial tests Interaction(a) and Interaction (b) since it produces a smaller p-value. Moreover, the significant factors or interaction are highlighted by the usual symbols used in the **summary** R function. If one does not specify that the exact distribution is desired, the **CSP.r** function obtains a Monte Carlo distribution of the test statistic with C = 1000 random permutations. The output would be as follows, if we set C = 5000:

```
> t<-CSP(y,x,C=5000)
> synchro.summary(t)
```

```
Monte Carlo Constrained Synchronized Permutation Testing
for Two-Way ANOVA
```

Source	T	Pr(>|T|)	
Factor A	347.3366	0.0324	*
Factor B	8.879206	0.4496	
Interaction (a)	188.9874	0.059	.
Interaction (b)	188.9874	0.0596	.
Interaction AB	377.9749	0.0324	*

```
Signif. codes:0 '***' 0.001 '**' 0.01 '*' 0.05 '.' 0.1' ' 1
```

```
Number of random permutations considered: 5000
Minimum achievable significance levels:
(a): 0.02857143        (b): 0.02857143
```

with some approximations in the p-value results. The F-test analysis gives $p_A = 0.0011$, $p_B = 0.4758$, and $p_{AB} = 0.0125$. The USP test with 5000 permutations gives $p_A = 2 \times 10^{-4}$, $p_B = 0.4615$, and $p_{AB} = 2 \times 10^{-4}$ (combined test). According to the F-test, factor A is very significant; note how both the CSP and the USP tests give a p-value for factor A that is equal to the minimum achievable significance level. In order to understand which are the levels that lead to the rejection of the null hypothesis, we can obtain synchronized permutation confidence intervals through the IC_USP.r function. This function computes the individual confidence interval for the difference of pairs of row means (default) or pairs of column means: In order to visualize the individual confidence intervals, it is first necessary to create a data frame (out, which will be the entry of the plot_hsu.r function) as follows:

```
> set.seed(101)
> source("IC_USP.r")
> n.comp<-choose(I,2)
> IC.A<-array(0,dim=c(n.comp,2))
> F<-array(0,dim=c(n.comp,2))
> m<-array(0,dim=c(n.comp,2))
> k<-1
> for(i in 1:(I-1)){
+ for(s in (i+1):I){
+ F[k,1]<-paste("m",i,sep="")
+ F[k,2]<-paste("m",s,sep="")
+ IC.A[k,]<-IC.USP(y,x,i,s)
```

```
+   m[k,1]<-mean(y[x[,1]==i])
+   m[k,2]<-mean(y[x[,1]==s])
+   k=k+1
+   }
+   }
> out.A<-data.frame(F1=F[,1],F2=F[,2],m1=m[,1],m2=m[,2],
+MSD=m[,1]-m[,2]-IC.A[,1])
>   out.A
  F1 F2         m1         m2 MSD
1 m1 m2 -1.3779272 0.1942948 0.83
2 m1 m3 -1.3779272 0.3353601 0.78
3 m2 m3  0.1942948 0.3353601 0.91
```

In order to view the individual confidence intervals for the differences $\alpha_i - \alpha_s$, $1 \leq i < s \leq 3$, type

```
colnames(IC)<-c("lower","upper")
rownames(IC)<-c("1-2","1-3","2-3")
> IC.A
          lower      upper
1-2 -2.402222 -0.7422220
1-3 -2.493287 -0.9332873
2-3 -1.051065  0.7689347
```

From these results we can conclude with a significance level of approximately 5%, that the null hypothesis $\alpha_2 = \alpha_3$ is not rejected.

The data frame out contains the informations required to create a plot like Figure 6.8: which pair of row or column means is considered, their values, and the half difference of the individual confidence intervals. The function IC_USP.r increases the value of MSD_{is} as in Algorithm 1 from 0.01 to max.delta/100 until the convergence criterion is satisfied. The entries conf.level and max.delta are set by default to 0.95, and 300, respectively. If the convergence criterion is not satisfied (the convergence criterion is $p_{is} < \alpha/2$, where p_{is} is the p-value obtained in Algorithm 1 as a function of MSD_{is}), the function will return a confidence interval in any case, even if the desired level of confidence has not been achieved. In this case, increase max.delta until the convergence is achieved. For instance, if we want to obtain the confidence interval for the difference of the effects of factor B (entry row.perm set equal to TRUE) at a 99% level with an $MSD_{is} \in [0.01, 0.5]$, we may type

```
> IC.USP(y,x,1,2,conf.level=0.99,row.perm=TRUE,max.delta=50)
1        1
2        0.503
3        0.478
4        0.453
5        0.451
```

```
. . . . . . . . .
45        0.124
46        0.097
47        0.126
48        0.106
49        0.099
       lower       upper
[1] -0.2416834   0.7383166
```

However, note that the confidence interval $[-0.2416834, 0.7383166]$ is actually an 80.2% confidence interval $(2 \times [0.5 - p_{is}])$. The numbers printed by the function are probabilities of rejecting the null hypothesis $\beta_1 = \beta_2$ as MSD_{is} increases (i.e., $p_{is} = \#\{T^*(MSD_{is}) \leq T^{obs}\}/B$ in Algorithm 1). By letting max.delta set to default, we obtain a 99% confidence interval equal to $[-0.7216834, 1.2183166]$. The convergence criterion is satisfied after 97 runs; however, note that the number of runs is itself a random variable. This is because the IC_USP.r applies Monte Carlo permutations, and therefore there is a monotone decreasing trend of the probabilities p_{is} from approximately 0.5 to the first value that satisfies the condition $p_{is} < \alpha/2$. For this reason, it is worth noting that the output of the function is a confidence interval with a level of confidence equal to or greater than $1 - \alpha$. With these settings, type

```
> set.seed(1)
> m1<-mean(y[x[,2]==1])
> m2<-mean(y[x[,2]==2])
> IC.B<-IC.USP(y,x,1,2,row.perm=TRUE)
> MSD=m1-m2-IC.B[1]
> out.B<-data.frame(F1="m1",F2="m2",m1=m1,m2=m2,MSD=MSD)
> out.B
  F1 F2         m1         m2  MSD
1 m1 m2 -0.1585991 -0.4069157 0.75
```

We are now ready to plot the ICs for factor A and factor B as follows:

```
> source("plot_hsu.r")
> plot.hsu(out.A,file="ICA",main="A",title="USP ICs",
+ dev="eps",measure=expression(sigma))
> plot.hsu(out.B,file="ICB",main="B",title="USP ICs",
+ dev="eps",measure=expression(sigma))
```

The plot_hsu.r function produces as output a file as in Figure 6.8. The entries of the functions are:

- a data frame containing the required information as the object out.A;
- the name of the output file, without extension; files can be viewed with the Ghostscript free software;
- the main title as usual in the R plots;
- a secondary title, in our example "USP ICs";

- the file extension (default is .eps, .pdf is another possible choice);
- a unit of measure, such as liters, meters, etc.

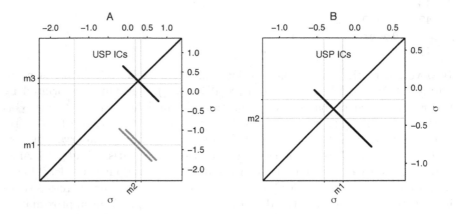

Fig. 6.8. Output of the function `plot_hsu.r` for both main factors.

It is possible to change the names of the levels by changing the levels of factor F. For instance, Figure 6.10 has been obtained by setting

```
> J<-3
> F<-array(0,dim=c((J*(J-1)/2),2))
> F[,1]<-c("Air","Air","Water")
> F[,2]<-c("Salt","Salt","Water")
```

The entries for the `plot_hsu.r` function are `main = "Individual CIs - USP"`, `title=""`, `measure = "Environment"`.

Figure 6.8 shows that there are no significant differences between the true effects of factor B (since the plotted confidence interval crosses the 45° line). When this happens, the segment reproducing the confidence interval is in bold black. The estimated means are $\bar{y}_{.1} = -0.1585991$ and $\bar{y}_{.2} = -0.4069157$. As regards factor A levels, there is no significant difference between α_2 and α_3 ($\bar{y}_{2.} = 0.1942948$ and $\bar{y}_{3.} = 0.3353601$), whereas α_1 is the effect that led to the rejection of the null hypothesis ($\bar{y}_{1.} = -1.3779272$). The CIs comparing α_1 to the other effects are highlighted in light grey because the related comparisons are significant. To conclude the example, let's compare the USP confidence intervals with those obtained with the well-known Tukey honest significant method

```
> fit<-aov(y ~ A*B)
> TukeyHSD(fit,"A")
  Tukey multiple comparisons of means
    95% family-wise confidence level
```

```
Fit: aov(formula = y ~ A * B)

$A
         diff        lwr      upr      p adj
2-1 1.5722220  0.5063812 2.638063 0.0038551
3-1 1.7132873  0.6474465 2.779128 0.0018395
3-2 0.1410653 -0.9247755 1.206906 0.9392540


> TukeyHSD(fit,"B")
  Tukey multiple comparisons of means
    95% family-wise confidence level

Fit: aov(formula = y ~ A * B)

$B
          diff        lwr       upr       p adj
2-1 -0.2483166 -0.9647045 0.4680712 0.4758417
```

As regards USP ICs:

```
> comp.A<-paste(F[,1],F[,2],sep="-")
> comp.B<-paste("m1","m2",sep="-")
> data.frame(comparison=comp.A,IC.A)
     comparison    lower      upper
1-2      m1-m2 -2.402222 -0.7422220
1-3      m1-m3 -2.493287 -0.9332873
2-3      m2-m3 -1.051065  0.7689347

> IC.B
    lower     upper
[1] 0.2283166 0.2683166
```

The individual confidence intervals obtained with the USPs are very similar to those obtained with Tukey's method, but note that no correction for multiplicity has been applied in the search of the intervals for $\alpha_i - \alpha_s$, $1 \le i < s \le 3$. To account for multiplicity (applying a Bonferroni correction), the conf.level entry of the IC_USP.r function must be set to $1 - \alpha/3$. The individual confidence intervals with a 95% confidence level accounting for multiplicity are shown in Table 6.5. Obviously, there is no need to correct the nominal confidence levels for $\beta_1 - \beta_2$. The number of permutations is set equal to 1000 by default. To increase the USP approximation, a larger number of permutations must be considered.

Both Tukey and USP CIs lead to the same conclusion: The effect that led to the rejection of H_{0A} is α_1 since the confidence intervals involving α_1 do not contain the zero.

Table 6.5. Individual USP CIs accounting for multiplicity (Bonferroni correction).

Comparison	Lower	Upper
$\alpha_1 - \alpha_2$	-2.50222	-0.642222
$\alpha_1 - \alpha_3$	-2.64328	-0.783287
$\alpha_2 - \alpha_3$	-1.20106	0.918934

6.7.2 Examples

The first example is Problem 5.11 from Montgomery (2001). This is a 2×3 design with $n = 3$. Factor A is the position of the furnace (position 1 or 2) and factor B is the firing temperature (800 °C, 825 °C, and 850 °C). The aim of the experiment was to determine how the main factors affected the density of a baked carbon anode. Table 6.6 shows the results obtained with the parametric two-way ANOVA and the synchronized permutation tests. Here n is too small to apply the CSPs since the number of distinct values of the test statistic is only ten. Recall that the minimum achievable significance level equals the inverse of the cardinality of distinct values of the CSP test statistic, which is $C_{CSP}/2$. For the USPs, instead the number of distinct values of the test statistic is 730 when the intermediate statistics are based on factor A, and 531442 when the intermediate statistics are based on factor B. Hence the test statistics are sensitive enough to be compared with a continuous distribution such as the F-statistic. In fact, the p-values obtained with the two-way ANOVA are very close to those obtained with the USPs. Instead, exact CSPs reach the minimum attainable p-value (which is $1/10$) for both main factors, which are strongly significant. As far as the interaction is concerned, we report two possible test statistics in the synchronized permutation strategy, depending on which main factor the intermediate statistic is based on. In Figure 6.9, the nonparametric confidence intervals at a 95% level are drawn for both main factor effects. From Figure 6.9 we can conclude that the effects of the two levels of factor A ($\bar{y}_{1..} = 729.9, \bar{y}_{2..} = 690$) are significantly different. Instead, the effect of levels 1 and 3 ($\bar{y}_{.1.} = 552.3, \bar{y}_{.3.} = 543.5$) of factor B cannot be considered to be significantly different, whereas the effect of level 2 ($\bar{y}_{.2.} = 1034$) is different from both of them.

The second example considered is Problem 5.22 from Montgomery (2001). Here there is an investigation on the effects of cycling loading and environmental conditions on fatigue crack growth at a constant 22 MPa stress for a particular material. This is a 3^2 factorial experiment with $n = 4$ replicates. Factor A is "frequency" (10, 1, and 0.1) and factor B is "environment" (air, water, and salt). The results for parametric two-way ANOVA and for the synchronized permutation tests are reported in Table 6.7. The p-values in the USP test column are all equal to zero because reasonably the Monte Carlo USP algorithm did not produce the observed value ($P[\nu = 0] = 7.87712E - 14$). We obtained these results with 10,000 random USPs.

Table 6.6. ANOVA and CSP/USP results for Problem 5.11 from Montgomery (2001).

		Two-Way ANOVA		
Source	D. F.	Sum of Squares	F-value	Pr > F
A	1	795.56	17.331	0.002
B	2	157557.01	1144.126	0.000
AB	2	272.7	0.99	0.427
Error	12	5370.66	-	-
		Synchronized Permutation Tests		
Source	Intermediate T	T-value	Pr > T_{USP}	Pr > T_{CSP}
A	$^{a}T_{is\mid j}$	128881	0.0011	0.1000
B	$^{b}T_{jh\mid i}$	17016158	0.0001	0.1000
AB	$^{a}T_{is\mid j}$	14726	0.4569	0.6000
AB	$^{b}T_{jh\mid i}$	14726	0.4732	0.6000

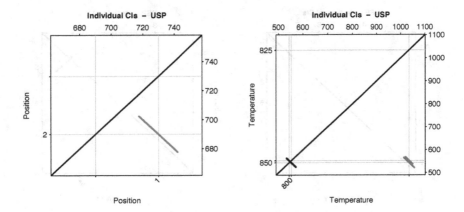

Fig. 6.9. Problem 5.11 from Montgomery (2001). 2×3 design with three replicates. Confidence intervals for factors "Position" and "Temperature".

Note that all the p-values in the CSP column are equal to the minimum attainable significance level, which is $1/35 = 0.02857$. Figure 6.10 shows the confidence intervals for both main factors in the proposed example. All the effects of factor A ($\bar{y}_{1..} = 2.114, \bar{y}_{2..} = 3.109, \bar{y}_{3..} = 7.660$) can be deduced to be significantly different. The effects of levels 2 and 3 of factor B ($\bar{y}_{.2.} = 5.392, \bar{y}_{.3.} = 5.077$) are significantly different from the effect of level 1 ($\bar{y}_{.1.} = 2.414$). Data are expressed in crack growth rates. The CIs were obtained at a confidence level of 95%.

Table 6.7. ANOVA and CSP/USP results for Problem 5.22 from Montgomery (2001).

Source	D. F.	Sum of Squares	F-Value	Pr > F
		Two-Way ANOVA		
A	2	17.49	104.94	0.000
B	2	5.35	32.12	0.000
AB	4	25.8	25.8	0.000
Error	27	5.42	-	-

Source	Intermediate T	T-value	Pr > T_{USP}	Pr > T_{CSP}
		Synchronized Permutation Tests		
A	${}^{a}T_{is\mid j}$	7556.14	0.000	0.02857
B	${}^{b}T_{jh\mid i}$	2313.07	0.000	0.02857
AB	${}^{a}T_{is\mid j}$	3670.76	0.000	0.02857
AB	${}^{b}T_{jh\mid i}$	3670.76	0.000	0.02857

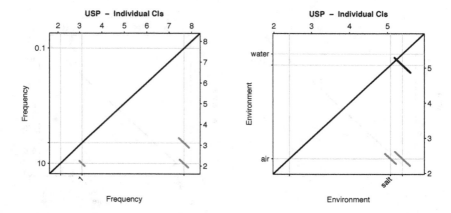

Fig. 6.10. Problem 5.22 from Montgomery (2001). 3×3 design with four replicates. Confidence intervals for factors "Frequency" and "Environment".

6.8 Further Developments

Synchronized permutation tests can also be applied to unbalanced design and two-way MANOVA. In the first case, some weights in the test statistics are required in order to obtain separate tests; in the latter, the nonparametric combination methodology applies. This last section is dedicated to these kinds of analyses.

6.8.1 Unbalanced Two-Way ANOVA Designs

In what follows we will refer to a 2×2 ANOVA unbalanced design. Let n_{11}, n_{12}, n_{21}, and n_{22} be respectively the sample sizes of blocks A_1B_1, A_1B_2, A_2B_1, and A_2B_2. It is still possible to apply synchronized permutations between each pair of blocks, but clearly here:

$$0 \leq \nu \leq \min\{n_{11}, n_{12}, n_{21}, n_{22}\}.$$

The idea is to weight the partial statistics in order to eliminate the confounding effects. For instance, the partial statistics to test for factor A become

$$^aT^*_{12|1} = w_{11} \sum_{k=1}^{n_{11}} y^*_{11k} - w_{21} \sum_{k=1}^{n_{21}} y^*_{21k} \tag{6.14}$$

$$^aT^*_{12|2} = w_{12} \sum_{k=1}^{n_{12}} y^*_{12k} - w_{22} \sum_{k=1}^{n_{22}} y^*_{22k} \tag{6.15}$$

and the weight associated with each block w_{ij}, $i, j = 1, 2$ must be determined in order to obtain separate tests. To identify the weights, let's consider the synchronized permutation structure of $^aT^*_{12|1}$ when exchanging ν units between blocks A_1B_1 and A_1B_2, and between A_2B_1 and A_2B_2. With the same notation as in Section 6.2,

$$\begin{aligned}
^aT^*_{12|1} &= w_{11}\left[(n_{11} - \nu)(\mu + \alpha_1 + \beta_1 + \alpha\beta_{11}) + \nu(\mu + \alpha_2 + \beta_1 + \alpha\beta_{21})\right] \\
&\quad - w_{21}\left[(n_{21} - \nu)(\mu + \alpha_2 + \beta_1 + \alpha\beta_{21}) + \nu(\mu + \alpha_1 + \beta_1 + \alpha\beta_{11})\right] + \hat{\varepsilon}^*_{12|1} \\
&= \left[w_{11}(n_{11} - \nu) - \nu w_{21}\right](\mu + \alpha_1 + \beta_1 + \alpha\beta_{11}) \\
&\quad - \left[w_{21}(n_{21} - \nu) - \nu w_{11}\right](\mu + \alpha_2 + \beta_1 + \alpha\beta_{21}) + \hat{\varepsilon}^*_{12|1}.
\end{aligned}$$

In order to eliminate the confounding effects μ and β_1, w_{11} and w_{21} must satisfy the condition

$$w_{11}(n_{11} - \nu) - \nu w_{21} = w_{21}(n_{21} - \nu) - \nu w_{11},$$

which leads to the condition $w_{11}n_{11} = w_{21}n_{21}$. Letting $w_{11} = w_{21}n_{21}n_{11}^{-1}$, $^aT^*_{12|1}$ becomes

$$^aT^*_{12|1} = w_{21}\left[n_{21} - \nu\frac{n_{11} + n_{21}}{n_{11}}\right](\alpha_1 - \alpha_2 + \alpha\beta_{11} - \alpha\beta_{21}) + \hat{\varepsilon}^*_{12|1}.$$

On the other hand, we have $w_{12}n_{12} = w_{22}n_{22}$ and

$$^aT^*_{12|2} = w_{22}\left[n_{22} - \nu\frac{n_{12} + n_{22}}{n_{12}}\right](\alpha_1 - \alpha_2 + \alpha\beta_{12} - \alpha\beta_{22}) + \hat{\varepsilon}^*_{12|2}.$$

In order to eliminate the interaction effect, by applying the side condition (6.1), w_{21} and w_{22} must satisfy

$$w_{21}\left[n_{21} - \nu\frac{n_{11} + n_{21}}{n_{11}}\right] = w_{22}\left[n_{22} - \nu\frac{n_{12} + n_{22}}{n_{12}}\right],$$

which leads to the relation

$$\omega_{21} = \omega_{22} \frac{n_{12}n_{22} - \nu(n_{12} + n_{22})\, n_{11}}{n_{11}n_{21} - \nu(n_{11} + n_{21})\, n_{12}}.$$

Since the system of the equations above is singular, without loss of generality, we let $\omega_{22} = [n_{11}n_{21} - \nu(n_{11} + n_{21})]n_{12}$ and $n_{21} = \min(n_{11}, n_{12}, n_{21}, n_{22})$. In this way, we are able to specify the weights in equations (6.14) and (6.15):

$$\omega_{11} = [n_{12}n_{22} - \nu(n_{12} + n_{22})]n_{21}$$
$$\omega_{21} = [n_{12}n_{22} - \nu(n_{12} + n_{22})]n_{11}$$
$$\omega_{12} = [n_{11}n_{21} - \nu(n_{11} + n_{21})]n_{22}$$
$$\omega_{22} = [n_{11}n_{21} - \nu(n_{11} + n_{21})]n_{12}$$

Note that ν must satisfy the condition

$$\nu \neq \frac{n_{11}n_{21}}{n_{11} + n_{21}}.$$

Finally, just apply the weights to the partial statistics of the CSP or USP test, depending on the available number sizes and the restriction on ν. The weights should be specified from the smallest n_{ij}.

6.8.2 Two-Way MANOVA

Let us now consider a two-way MANOVA layout where a Q-dimensional random variable is observed on a sample of nIJ units from a replicated two-factor experimental design. In the previous sections, we have provided a solution for the univariate case. The multivariate case requires the nonparametric combination methodology introduced in Chapter 1. Let

$$\mathbf{Y}_{ijk} = [_1Y_{ijk}, _2Y_{ijk}, \ldots, _QY_{ijk}]$$

be the (row) vector of the multivariate response associated to the kth experimental units from block A_iB_j with k, I, and J as in Section 6.1. Also assume that each component of the multivariate response follows an addictive model:

$$_qY_{ijk} = {}_q\mu + {}_q\alpha_i + {}_q\beta_j + {}_q\alpha\beta_{ij} + {}_q\varepsilon_{ijk} \qquad q = 1, \ldots, Q.$$

The components of the multivariate response $\mathbf{Y}_{ijk}$ could be either dependent or independent; there is no need to specify (or to estimate) the covariance matrix. It sometimes could be useful to assess whether a two-factor experiment has at least some significant effects on a set of Q variables. Let us refer, for instance, to factor A. Let

$$_qH_{0A} : {}_q\alpha_1 = {}_q\alpha_2 = \ldots, {}_q\alpha_I \qquad q = 1, \ldots, Q$$

be the (partial) null hypothesis for the effects of factor A on the qth variable. If factor A has no real effect on any of the Q variables, then this can be written as a global null hypothesis,

$$H_{0A}^{G} = \cap_{q=1}^{Q} {}_{q}H_{0A},$$

where H_{0A}^{G} is true if and only if all the considered partial null hypotheses are true. On the other hand, the alternative is true if at least one partial null hypothesis is false. We may write that as follows:

$$H_{1A}^{G} = \cup_{q=1}^{Q} {}_{q}H_{1A}.$$

Now, in order to obtain a global test to assess H_{0A}^{G}, we first perform Q partial tests on the effects of factor A on the qth component of the response and then combine the partial tests through the nonparametric combination. To do this, proceed as follows:

1. For $q = 1, \ldots, Q$, repeat:
   a) Obtain the synchronized permutation distribution (CSP or USP) of the test statistic for the effects of factor A on the qth variable,

   $$_{q}^{a}\mathbf{T}_{A}^{*} = [{}_{q}^{a}T_{A}, {}_{q}^{a}T_{A}^{*(1)}, {}_{q}^{a}T_{A}^{*(2)}, \ldots, {}_{q}^{a}T_{A}^{*(b)}, \ldots, {}_{q}^{a}T_{A}^{*(B)}]',$$

   where ${}_{q}^{a}T_{A}$ is the observed value of the test statistic and ${}_{q}^{a}T_{A}^{*(b)}, b = 1, \ldots, B$, is the generic element of its distribution.
   b) Obtain the vector of p-values

   $$_{q}\mathbf{P}_{A}^{*} = [{}_{q}p_{A}, {}_{q}p_{A}^{*(1)}, {}_{q}p_{A}^{*(2)}, \ldots, {}_{q}p_{A}^{*(b)}, \ldots, {}_{q}p_{A}^{*(B)}]',$$

   where ${}_{q}p_{A}$ is the observed p-value of the effects of factor A and ${}_{q}p_{A}^{*(b)}, b = 1, \ldots, B$ is the p-value that would be obtained if ${}_{q}^{a}T_{A}^{*(b)}$ were the observed value of the test statistic,

   $$_{q}p_{A}^{*(b)} = \frac{1}{B} \sum_{j=1}^{B} I\left({}_{q}^{a}T_{A}^{*(j)} \geq {}_{q}^{a}T_{A}^{*(b)}\right).$$

2. Consider the matrix $\mathbf{P}_{A}^{*}$ whose columns are made by each vector ${}_{q}\mathbf{P}_{A}^{*}$:

   $$\mathbf{P}_{A}^{*} = [{}_{1}\mathbf{P}_{A}^{*}, {}_{2}\mathbf{P}_{A}^{*}, \ldots, {}_{Q}\mathbf{P}_{A}^{*}].$$

   Note that the first row of $\mathbf{P}_{A}^{*}$ is made of the partial p-values for the effects of factor A on each variable.
3. Apply a suitable combining function $\psi(\cdot)$ to the rows of $\mathbf{P}_{A}^{*}$ and obtain the vector

   $$\boldsymbol{\Psi} = \left[\psi, \psi^{*(1)}, \psi^{*(2)}, \ldots, \psi^{*(b)}, \ldots, \psi^{*(B)}\right]',$$

   where $\psi = \psi({}_{1}p_{A}, {}_{2}p_{A}, \ldots, {}_{Q}p_{A})$, and $\psi^{*(b)} = \psi({}_{1}p_{A}^{*(b)}, {}_{2}p_{A}^{*(b)}, \ldots, {}_{Q}p_{A}^{*(b)})$.

4. If large values of ψ are significant, obtain the global p-value as:

$$p_A^G = \frac{1}{B} \sum_{b=1}^{B} I\left(\psi^{*(b)} \geq \psi\right).$$

5. If $p_A^G > \alpha$, where α is the desired type I error rate, accept H_{0A}^G; otherwise reject H_{0A}^G.

If H_{0A}^G is rejected, one can evaluate on which variable(s) the effects of factor A are significant by looking at the first row of $\mathbf{P_A^*}$ and by applying suitable corrections for multiplicity. If the components of the response are known to be independent, one can perfom points 1(a) and 1(b) of the algorithm above independently. Otherwise, a further synchronization among the Q components of the response is required. To better understand this case, consider the unit-by-unit representation of a balanced 2×2 MANOVA design as in Table 6.8. If CSPs are applied, $\nu = 1$, and the units to be swapped whitin pairs of blocks are the first ones of each block, then we should swap the rows indicated by the symbol "$\leftarrow$" with each other and the rows indicated by the symbol "$\Leftarrow$" with each other. That is, the swapping has to be made while maintaining the inner dependence among the components of the response in each unit.

Table 6.8. Unit-by-unit MANOVA design representation.

Indexes			Variables			
i j	k		$_1\mathbf{Y}$	$_2\mathbf{Y}$	$\cdots$	$_Q\mathbf{Y}$
1 1	1		$_1y_{111}$	$_2y_{111}$	$\cdots$	$_Qy_{111}$ $\leftarrow$
$\vdots$ $\vdots$	$\vdots$		$\vdots$	$\vdots$	$\vdots$	$\vdots$
1 1	n		$_1y_{11n}$	$_2y_{11n}$	$\cdots$	$_Qy_{11n}$
1 2	1		$_1y_{121}$	$_2y_{121}$	$\cdots$	$_Qy_{121}$ $\Leftarrow$
$\vdots$ $\vdots$	$\vdots$		$\vdots$	$\vdots$	$\vdots$	$\vdots$
1 2	n		$_1y_{12n}$	$_2y_{12n}$	$\cdots$	$_Qy_{12n}$
2 1	1		$_1y_{211}$	$_2y_{211}$	$\cdots$	$_Qy_{211}$ $\leftarrow$
$\vdots$ $\vdots$	$\vdots$		$\vdots$	$\vdots$	$\vdots$	$\vdots$
2 1	n		$_1y_{21n}$	$_2y_{21n}$	$\cdots$	$_Qy_{21n}$
2 2	1		$_1y_{221}$	$_2y_{221}$	$\cdots$	$_Qy_{221}$ $\Leftarrow$
$\vdots$ $\vdots$	$\vdots$		$\vdots$	$\vdots$	$\vdots$	$\vdots$
2 2	n		$_1y_{22n}$	$_2y_{22n}$	$\cdots$	$_Qy_{22n}$

7

Permutation Tests for Unreplicated Factorial Designs

In experimental design, wide use is made of screening comparative experiments to compare the effects of treatments on a given number of experimental units. In this type of experimentation, designing the experiment in such a way as to obtain all pertinent information in the related field of research is of fundamental importance, especially for costly or complicated experiments. Screening designs very often represent the first approach to experimental situations in which many explanatory factors are available and we are interested in establishing which ones are significant. In this field, the two-level factorial designs represent an instrument that is easy to use and interpret. When considering complete unreplicated designs, it is not possible to obtain an estimate of the variance of the errors, and therefore the usual inferential techniques, aimed at identifying the significantly active factors, are unsuitable. Various solutions to this problem have been proposed in the literature. Some, such as the *normal plot* (Daniel, 1959), prove not to be very objective; others presuppose assumptions, such as normality of errors, even when little or nothing is known about the phenomenon being analyzed. Hamada and Balakrishnan (1998) propose a review of existing parametric and nonparametric tests conceived for unreplicated two-level factorial designs.

It is worth noting that the first proposal of a permutation approach for testing active effects in factorial designs appeared in Loughin and Noble (1997); Pesarin and Salmaso (2002) proposed an exact permutation test for the K largest effects; finally Basso and Salmaso (2006) introduced two versions of a permutation test to control the individual error rate (IER) and the experimental error rate (EER). These kinds of errors will be detailed in what follows. The test controlling IER, called T_P, does not need to be calibrated and allows testing for all $2^K - 1$ effects in the design. It is described in Section 7.4. A sequential version of the same test, called *step-up T_P*, allows us to control EER but requires a proper calibration. The updated version of the *step-up T_P* test is described in Section 7.5, and a simple way to calibrate it in order to control the EER is introduced in Subsection 7.5.1. This test allows us to test for the $2^K - 2$ largest effects.

D. Basso et al., *Permutation Tests for Stochastic Ordering and ANOVA*, Lecture Notes in Statistics, 194, DOI 10.1007/978-0-387-85956-9_7,
© Springer Science+Business Media, LLC 2009

It does not seem possible to obtain exact permutation tests for all the effects in a 2^K unreplicated factorial design, as the requirement of exchangeability of the responses is not generally satisfied here.

One of the desirable properties for a testing procedure is the robustness of the test in the conclusions; that is, the ability of the test to identify active effects even without any assumption on the distribution of the experimental errors (Pesarin, 2001).

7.1 Brief Introduction to Unreplicated 2^K Full Factorial Designs

Unreplicated 2^K full factorial designs are experimental designs where the effects of K main factors on the response are investigated. The main factors are usually identified by capital letters A, B, etc., and assume only two levels: the low level (coded by -1) and the high level (coded by $+1$). Moreover, the interaction effects related to a pair of main factors (two-order interactions) or to $C > 2$ main factors (higher-order interactions) are considered. Usually the interactions are denoted by the capital letters of the main factors they involve. From now on, we will not distinguish main factors from interactions and simply call them all (experimental) "factors". In a full factorial design, all the possible treatments (i.e., combinations of the main factor levels) are investigated. In unreplicated designs, a single run is considered for each treatment; therefore in unreplicated 2^K full factorial designs, the number of responses is equal to $n = 2^K$. The design matrix of an unreplicated 2^3 full factorial design is as follows:

$$\mathbf{X} = \begin{bmatrix} \mu & A & B & C & AB & AC & BC & ABC \\ 1 & 1 & 1 & 1 & 1 & 1 & 1 & 1 \\ 1 & 1 & 1 & -1 & 1 & -1 & -1 & -1 \\ 1 & 1 & -1 & 1 & -1 & 1 & -1 & -1 \\ 1 & 1 & -1 & -1 & -1 & -1 & 1 & 1 \\ 1 & -1 & 1 & 1 & -1 & -1 & 1 & -1 \\ 1 & -1 & 1 & -1 & -1 & 1 & -1 & 1 \\ 1 & -1 & -1 & 1 & 1 & -1 & -1 & 1 \\ 1 & -1 & -1 & -1 & 1 & 1 & 1 & -1 \end{bmatrix}.$$

Each row of $\mathbf{X}$ corresponds to a different combination of the ± 1 main factor levels (i.e., a treatment). The first column of $\mathbf{X}$ refers to the intercept of the linear model, and the remaining columns are related to the experimental factors. The design matrix $\mathbf{X}$ is also known as a *Hadamard* matrix. A Hadamard matrix is a square matrix whose columns are orthogonal and that satisfies the conditions $\mathbf{X}'\mathbf{X} = \mathbf{X}\mathbf{X}' = n\mathbf{I}_n$, where $\mathbf{I}_n$ is the identity matrix of order n.

The aim of the researcher is to assess the null hypothesis on the effects of all factors. The effect of a factor is defined as the difference in the response means corresponding to its +1 and −1 levels. A linear model is usually applied to fit the response

$$\mathbf{Y} = \mathbf{X}\boldsymbol{\beta} + \boldsymbol{\varepsilon}, \tag{7.1}$$

where $\mathbf{Y}$ is an $n \times 1$ vector of responses, $\mathbf{X}$ is the $n \times p$ design matrix, $\boldsymbol{\beta}$ is a $p \times 1$ vector of parameters (i.e., $\boldsymbol{\beta} = [\beta_0, \beta_1, \ldots, \beta_{p-1}]'$), and $\boldsymbol{\varepsilon}$ is an $n \times 1$ vector of exchangeable experimental errors from an unspecified distribution with zero mean and finite variance σ^2.

Although in unreplicated 2^K full factorial designs $n = p$, we will refer to n as the number of observations in the response and to p as the number of parameters in the linear model. We will denote the intercept as β_0, the effect corresponding to factor A as β_1, and so on. The interaction involving all main factors will be denoted as β_{p-1}.

The estimates of the effects are obtained by applying the ordinary least squares estimation to the set of observed data $\mathbf{y}$

$$\hat{\boldsymbol{\beta}} = [\mathbf{X}'\mathbf{X}]^{-1}\mathbf{X}'\mathbf{y} = [\hat{\beta}_0, \hat{\beta}_1, \ldots, \hat{\beta}_{p-1}]',$$

where $\hat{\beta}_0 = \bar{y}$ is the sample mean of observed responses and $\hat{\beta}_j$, $j = 1, \ldots, p-1$, are the estimates of the effects. Since $n = p$, the linear model exactly fits the response, so the residual deviance of the model is null (and the total deviance equals the explained deviance). The explained deviance of the model is

$$SSE_p = \sum_{i=1}^{n} [\hat{y}_i - \bar{y}]^2 = \sum_{i=1}^{n} \left[\sum_{j=0}^{p-1} \hat{\beta}_j x_{ij} - \hat{\beta}_0 \right]^2$$

$$= \sum_{i=1}^{n} \sum_{j=1}^{p-1} \hat{\beta}_j^2 x_{ij}^2 + \sum_{i=1}^{n} \sum_{j=1}^{p-1} \sum_{l \neq j} \hat{\beta}_j \hat{\beta}_l x_{ij} x_{il}$$

$$= n \sum_{j=1}^{p-1} \hat{\beta}_j^2,$$

where the last result is due to the fact that the columns of $\mathbf{X}$ are orthogonal and that they are all made of ± 1 elements. Note that the total deviance can thus be decomposed into the sum of $p - 1$ uncorrelated random variables $\hat{\beta}_1^2, \hat{\beta}_2^2, \ldots, \hat{\beta}_{p-1}^2$. In fact, from the theory on linear models, we have $V(\hat{\boldsymbol{\beta}}) = \sigma^2 [\mathbf{X}'\mathbf{X}]^{-1} = n^{-1}\sigma^2 \mathbf{I}_n$.

The decomposition of the total deviance will be the starting point to define the test statistic in Section 7.3, and it also plays an important role in the definition of Loughin and Noble's (1997) test.

Let us briefly discuss the hypotheses under testing. Usually, the experimenter's major interest is in testing separately for main effects and for interactions. Hence, there are $p - 1$ null hypotheses that are of interest: $H_{0\beta_1} : \{\beta_1 = 0\}$, $H_{0\beta_2} : \{\beta_2 = 0\}$, ..., $H_{0\beta_{p-1}} : \{\beta_{p-1} = 0\}$.

Since $p - 1$ tests are to be done simultaneously, two kinds of errors may arise: (i) the *individual error rate* (IER), which is defined as the probability of incorrectly declaring one effect active (i.e., significant) irrespective of what happens with the other effects; and (ii) the *experiment-wise error rate* (EER), which is the probability of incorrectly declaring at least one effect active when they are all inactive. The IER can be computed as the average of each single factor rejection rate or can be summarized as $\sum_{i=1}^{p-1} i\pi_i$, where i is the number of factors declared active and π_i is the probability of declaring i active effects. The EER is given by $1 - \pi_0$ in accordance with the last notation. Which error should be controlled depends on the objectives of the experimenter. It does not seem possible to control both kinds of errors. As reported in Hamada and Balakrishnan (1998), the existing procedures for unreplicated 2^K factorials controlling EER are in general sequential procedures.

Since there are no degrees of freedom left to estimate the error variance (and since the residual variance is null), the parametric tests, which generally assume the normality of the error distribution and require an unbiased estimate of the error variance, cannot be applied here. Lenth (1989) proposed a parametric test for the unreplicated 2^K full factorial design by providing a robust pseudo-estimate of the error variance based on the number of negligible effects. This solution requires the assumption of *effect sparsity*, i.e., the estimate of the error variance requires some effect to be inactive. Although this assumption may be reasonable, Lenth's test cannot be considered completely reliable since the procedure initially involves all the estimates of the effects (although the median is taken as the estimator). The more active effects are present, the more the estimated error variance is inflated.

7.2 Loughin and Noble's Test

Loughin and Noble (1997) introduced a permutation test on effects for un-replicated factorials that represents the first permutation approach to this problem. The requirement of a permutation test is the exchangeability of the elements of the response, which does not generally hold since each observation is associated to a different treatment, and therefore the observations are not identically distributed (at least their expected values depend on the treatment received).

Thus, the authors introduced a sequence of null hypotheses in order to test for as many effects as possible according to the requirement of exchangeability of the response elements. To do so, they introduced the residualization of the response with respect to the previously tested effects in order to eliminate their influence on the response.

Loughin and Noble's test is a sequential procedure we are now going to illustrate. Let $|\hat{\beta}_{(1)}| \geq |\hat{\beta}_{(2)}| \geq \cdots \geq |\hat{\beta}_{(p-1)}|$ be the ordered absolute estimates of the effects. Since, unconditionally, $E[\hat{\beta}_j = \beta_j]$, the observed order reasonably reflects the true ordering of the absolute effects. Suppose we want to test for the null hypothesis $H_{0(1)} : \beta_{(1)} = 0$. Clearly, if $H_{0(1)}$ is true, then all the remaining effects are inactive too, and therefore the relationship between elements of $\mathbf{y}$ and $\mathbf{X}$ is determined purely by the randomization of the factor levels to the runs of the experiment. In other words, the elements of the response are exchangeable under $H_{0(1)}$ since $H_{0(1)}$ is a global null hypothesis involving all the effects. This allows us to obtain a reference distribution for the test statistic

$$W_{(1)} = |\hat{\beta}_{(1)}|, \tag{7.2}$$

by computing the test statistic (7.2) for every possible permutation of the response elements $\mathbf{y}^*$. Let

$$G(w^*_{(1)}|\mathbf{y}) = P[W^*_{(1)} \leq w^*_{(1)}|\mathbf{y}]$$

be the permutation cumulative distribution function of the random variable $W^*_{(1)} = \hat{\beta}^*_{(1)}|$. The $H_{0(1)}$ will be rejected at a significance level α if

$$1 - G(W_{(1)}|\mathbf{y}) < \alpha.$$

The permutation test on the largest effect is exact since $G(w^*_{(1)}|\mathbf{y})$ is the conditional distribution of $W^*_{(1)}$ under $H_{0(1)}$. Note that if $H_{0(1)}$ is not rejected, the testing procedure should stop and conclude that all the effects are inactive. If $H_{0(1)}$ is rejected, the elements of the response are not exchangeable since their unconditional expected value (may) depend on $\beta_{(1)} \neq 0$. Therefore, in order to test for the remaining $p - 2$ effects, the authors suggest applying the residualization of the response with respect to the largest estimated effect $\hat{\beta}_{(1)}$. That is, they suggest considering

$$\tilde{\mathbf{y}}_2 = \mathbf{y} - \hat{\beta}_{(1)}\mathbf{x}_{(1)}$$

as the new vector of observations, where $\mathbf{x}_{(1)}$ is the column of the design matrix that generated $\hat{\beta}_{(1)}$. Therefore, the OLS estimates of the effects from $\tilde{\mathbf{y}}_2$ become:

$$\tilde{\beta} = [\mathbf{X}'\mathbf{X}]^{-1}\mathbf{X}'\tilde{\mathbf{y}}_2 = [\mathbf{X}'\mathbf{X}]^{-1}\mathbf{X}'\mathbf{y} - \hat{\beta}_{(1)}[\mathbf{X}'\mathbf{X}]^{-1}\mathbf{X}'\mathbf{x}_{(1)}$$
$$= \hat{\beta} - \hat{\beta}_{(1)}\mathbf{u}_{(1)},$$

where $\mathbf{u}_{(1)} = [I(\hat{\beta}_1 = \hat{\beta}_{(1)}), I(\hat{\beta}_2 = \hat{\beta}_{(1)}), \ldots, I(\hat{\beta}_{p-1} = \hat{\beta}_{(1)})]'$ and $I(\cdot)$ is the indicator function. This result is due to the orthogonality of the columns of $\mathbf{X}$, so $\mathbf{x}'_j\mathbf{x}_{(1)} = 0$ if $\mathbf{x}_j \neq \mathbf{x}_{(1)}$ and $\mathbf{x}'_j\mathbf{x}_{(1)} = 1$ if $\mathbf{x}_j = \mathbf{x}_{(1)}$, $j = 1, \ldots, p$. The new vector of ordered absolute estimates of the effects is then

$$\tilde{\beta} = [\tilde{\beta}_{(1)} = \hat{\beta}_{(2)}, \tilde{\beta}_{(2)} = \hat{\beta}_{(3)}, \ldots, \tilde{\beta}_{(p-2)} = \hat{\beta}_{(p-1)}, \tilde{\beta}_{(p-1)} = 0]'.$$

Therefore $\hat{\beta}_{(2)}$ is now the largest observed estimate of the effects, and it is possible to assess the null hypothesis $H_{0(2)} : \beta_{(2)} = 0$ by choosing as a test statistic

$$W_{(2)} = |\tilde{\beta}_{(1)}|.$$

The elements of the new response $\tilde{\mathbf{y}}_2$ are still not exchangeable, though they can be considered approximately exchangeable (we should apply the residualization of the response with respect to the true effect $\beta_{(1)}$ in order to preserve the exchangeability under $H_{0(2)}$). Note that the residualization made on $\mathbf{y}$ with respect to $\hat{\beta}_{(1)}$ (which is a linear combination of the elements of $\mathbf{y}$) introduces correlations among the elements of $\tilde{\mathbf{y}}_2$, so they are no longer independent. Despite that, a permutation reference distribution for $W_{(2)}$ can be obtained by computing $W_{(2)}^* = |\tilde{\beta}_{(1)}^*|$, the largest absolute estimate of the effects obtained from every possible permutation of the vector $\tilde{\mathbf{y}}_2$. On the one hand, the explained deviance of the linear model when the estimates of the effects are obtined from $\tilde{\mathbf{y}}_2$ is

$$SSE(\tilde{\mathbf{y}_2}) = \sum_i [\tilde{y}_i - \bar{y}]^2 = \sum_i [y_i - \hat{\beta}_{(1)} x_{i(1)} - \hat{\beta}_0]^2$$

$$= \sum_i \left[\sum_{j=1}^{p-1} \hat{\beta}_{(j)} x_{i(j)} - \hat{\beta}_{(1)} x_{i(1)} \right]^2$$

$$= n \sum_{j=2}^{p-1} \hat{\beta}_{(j)}^2.$$

On the other hand, if we consider a random permutation of the residualized response $\mathbf{y}_2^*$, we have

$$SSE(\tilde{\mathbf{y}_2^*}) = \sum_i [\tilde{y}_i^* - \bar{y}]^2 = n \sum_{j=1}^{p-1} \tilde{\beta}_j^{*2},$$

where $\tilde{\beta}_j^{*2}$ is the estimate of the jth effect obtained from $\tilde{\mathbf{y}}_2^*$. Since $\tilde{\mathbf{y}}_2^*$ is a random permutation of $\tilde{\mathbf{y}}_2$, we have $SSE(\tilde{\mathbf{y}}_2) = SSE(\tilde{\mathbf{y}}_2)^*$. Therefore, the total variability is decomposed into $p - 2$ nonnull estimates from $\tilde{\mathbf{y}}_2$, but at each permutation it is decomposed into $p - 1$ nonnull estimates from $\tilde{\mathbf{y}}_2^*$. Note that this does not always happen: Given that permuting the response is equivalent (gives the same estimates) to permuting the rows of the design matrix and keeping the response fixed, we have that $\tilde{\beta}_j^* = 0$ whenever the jth column of $\mathbf{X}^*$ is equal to $\mathbf{x}_{(1)}$. Unconditionally

$$E[SSE(\tilde{\mathbf{y}_2})] = n \sum_{j=2}^{p-1} E\left[\hat{\beta}_{(j)}^2 \right]$$

$$= n \sum_{j=2}^{p-1} \left[\frac{\sigma^2}{n} + \beta_{(j)}^2 \right]$$

$$= \sigma^2(p - 2) + \sum_{j=2}^{p-1} \beta_{(j)}^2.$$

Under $H_{0(2)}$, $E[SSE(\tilde{\mathbf{y}}_2)|H_{0(2)}] = \sigma^2(p-2)$. In a similar way we can prove that $E[SSE(\tilde{\mathbf{y}}_2^*)|H_{0(2)}] = \sigma^2(p-1)$, and this means that, on average, the permutation estimates are $(p-2)/(p-1)$ times smaller than the observed estimates. Loughin and Noble suggest applying an empirical correction to the permutation estimates, which should be computed as

$$\tilde{\gamma}_j^* = \left(\frac{p-1}{p-2}\right)^{1/2} \tilde{\beta}_j^* \qquad j = 1, \ldots, p-1. \qquad (7.3)$$

Finally, a reference distribution for $\hat{W}_2 = |\hat{\beta}|_{(2)}$, the largest of the OLS estimates from $\tilde{\mathbf{y}}_2$, is obtained from the permutation distribution of $W_2^* = |\tilde{\gamma}^*|_{(1)} = \max_i |\tilde{\gamma}_i^*|$. The authors suggest a correction of the p-value for the second tested effect, claiming that $\hat{W}_2$ is the maximum over $p-2$ random variables, whereas W_2^* is the maximum over $p-1$ random variables, so that the p-value related to $\hat{W}_2$ should be computed as

$$P_2 = 1 - [G(\hat{W}_2|\tilde{\mathbf{y}}_2)]^{(p-2)/(p-1)}, \qquad (7.4)$$

where $G(W_2^*|\tilde{\mathbf{y}}_2)$ is the permutation c.d.f. of W_2^*. This correction is also empirical since whenever a column of $\mathbf{X}^*$ (a row permutation of $\mathbf{X}$) is equal to $\mathbf{x}_{(1)}$, W_2^* is the maximum over $p-2$ random variables as well. The remaining effects are tested similarly by replacing $p-2$ with $p-s$ in (7.3) and (7.4). The algorithm of the Loughin and Noble test can be written as follows:

1. Compute $\hat{\boldsymbol{\beta}} = [\hat{\beta}_1, \hat{\beta}_2, \ldots, \hat{\beta}_{p-1}]'$ from $\mathbf{y}$, and order the effects $|\hat{\beta}_{(1)}| \geq |\hat{\beta}_{(2)}| \geq \cdots \geq |\hat{\beta}_{(p-1)}|$.
2. At step $s = 1, \ldots, p-2$, let

$$\hat{W}_s = |\hat{\beta}_{(s)}|$$

   and obtain

$$\tilde{\mathbf{y}}_\mathbf{s} = \tilde{\mathbf{y}} - \hat{\beta}_{(1)}\mathbf{x}_{(1)} - \hat{\beta}_{(2)}\mathbf{x}_{(2)} - \cdots - \hat{\beta}_{(s-1)}\mathbf{x}_{(s-1)}$$

   (when $s = 1$, let $\tilde{\mathbf{y}}_1 = \mathbf{y}$).
3. Repeat B times:
   (a) Obtain $\tilde{\mathbf{y}}_\mathbf{s}^*$, a random permutation of $\tilde{\mathbf{y}}_\mathbf{s}$.
   (b) Compute $\tilde{\boldsymbol{\beta}}^* = [\tilde{\beta}_1^*, \tilde{\beta}_2^*, \ldots, \tilde{\beta}_{p-1}^*]'$ from $\tilde{\mathbf{y}}_\mathbf{s}^*$.
   (c) Let

$$W_s^* = \left(\frac{p-1}{p-s}\right)^{1/2} |\tilde{\beta}_{(1)}^*|.$$

4. Compute the p-value of the test as

$$p_{(s)} = 1 - \left[\frac{\#W_s^* \leq \hat{W}_s}{B}\right]^{(p-s)/(p-1)}.$$

This algorithm allows us to test for $p - 2$ effects. The smallest effect cannot be tested because $SSE(\tilde{\mathbf{y}}_{\mathbf{p-1}}) = n\hat{\beta}^2_{(p-1)}$ and $n\tilde{\beta}^{*2}_j \geq SSE(\tilde{\mathbf{y}}_{\mathbf{p-1}})/(p-1)$. Therefore $W^*_{p-1} = \sqrt{p-1}\tilde{\beta}^*_j \geq \hat{\beta}_{(p-1)} = \hat{W}_{p-1}$ and $p_{(p-1)} \equiv 1$.

The empirical corrections (7.3) and (7.4) may work well with the largest effects, but they are not reliable for the smaller effects. Each time the response is residualized, the chance of observing null permutation values increases and the permutation distribution degenerates; see Basso and Salmaso (2007) for details.

The presence of (many) potentially active effects may produce a "masking-effect", as pointed out in an artificial example from Loughin and Noble where data have been generated by considering four active effects of the same size and no errors in a 2^4 experiment. The p-values $p_{(1)}$-$p_{(4)}$ they obtained were equal to 0.9198, 0.5923, 0.1482, and 0.0019. Only the fourth ordered effect is really significant against the related null hypothesis, and this happens because, when $\beta_{(4)}$ is being tested, the previous active effects have already been removed. Thus, the (explained) variability due to the previously tested effect is no longer partitioned in the permutation values. Since no errors have been considered, $\hat{\beta}_j = \beta_j$ $j = 1, \ldots, p - 1$. Therefore four active effects of the same size are compared to four permutation distributions that satisfy $\tilde{W}^*_4 \overset{d}{<} \tilde{W}^*_3 \overset{d}{<} \tilde{W}^*_2 \overset{d}{<} \tilde{W}^*_1$. To avoid the masking effect, the authors suggested a step-up interpretation of the results (that is, from $p_{(p-2)}$ to $\ldots, p_{(1)}$). The step-up interpretation modifies the underlying null hypothesis in the sense that if the null hypothesis is rejected for one effect, then it must be rejected for all larger effects. This can be done by considering significant all the effects that are larger than the smallest effect for which $p_{(s)} \leq p_0$, where p_0 is a suitable critical value to give the procedure the desired type I error rate. The authors provided the critical p-values p_0 corresponding to various sizes of experimental designs and desired error rates (IER or EER). The critical p-values were estimated from an intensive simulation study, and we refer to Loughin and Noble (1997) for details.

7.3 The T_F Test

When fitting a linear model for an unreplicated 2^K full factorial design, the main problem is that the residual variance cannot be estimated because of the lack of degrees of freedom. The residual deviance of the linear model (7.1) is null and the explained deviance is partitioned into $p - 1$ uncorrelated random variables (the squared estimated effects). In the previous section, we described Loughin and Noble's testing procedure, which applies the residualization of the response with respect to the effects previously tested. If we consider the second step of their algorithm and focus on $\boldsymbol{\Sigma}(\tilde{\mathbf{y}}_2)$ (i.e., the unconditional covariance matrix of $\tilde{\mathbf{y}}_2$), we have

$$\Sigma(\tilde{\mathbf{y}}_2) = E[\mathbf{y} - \hat{\beta}_{(1)}\mathbf{x_1} - E(\mathbf{y} - \hat{\beta}_{(1)}\mathbf{x_1})][\mathbf{y} - \hat{\beta}_{(1)}\mathbf{x_1} - E(\mathbf{y} - \hat{\beta}_{(1)}\mathbf{x_1})]'$$
$$= E[\mathbf{y} - E(\mathbf{y})][\mathbf{y} - E(\mathbf{y})]' - E[\hat{\beta}_{(1)}^2 - E(\hat{\beta}_{(1)})]^2 \mathbf{x}_{(1)}\mathbf{x}_{(1)}'$$
$$= \sigma^2 \left[\mathbf{I_n} - \frac{1}{n}\mathbf{x}_{(1)}\mathbf{x}_{(1)}' \right].$$

By induction on $\tilde{\mathbf{y}}_s = \tilde{\mathbf{y}}_{s-1} - \hat{\beta}_{(s-1)}\mathbf{x}_{(s-1)}$, the covariance matrix of the response elements $\tilde{\mathbf{y}}_s$ at step s is:

$$\Sigma(\tilde{\mathbf{y}}_s) = \sigma^2 \left[\mathbf{I_n} - \frac{1}{n}\mathbf{x}_{(1)}\mathbf{x}_{(1)}' - \frac{1}{n}\mathbf{x}_{(2)}\mathbf{x}_{(2)}' - \cdots - \frac{1}{n}\mathbf{x}_{(s-1)}\mathbf{x}_{(s-1)}' \right].$$

A permutation test is based on the notion of exchangeability of random variables, which does not necessarily mean independence. However, it is hard to assume the exchangeability of the elements in $\tilde{\mathbf{y}}_s$ since, for instance, the pairwise correlations are different (the matrices $\mathbf{x}_{(s)}\mathbf{x}_{(s)}'$ have ± 1 elements). When the assumption of exchangeability does not hold, the permutation test becomes approximated, and as the correlation among the response elements "increases" with s, the tests on the smallest effects are not reliable. Note that the tests on the smallest effects play an important role in the step-up interpretation of the results, which is suggested by Loughin and Noble themselves. This consideration suggests that the residualization of the response is probably not the best way to follow in order to apply a permutation test. On the other hand, a step-up procedure can avoid the masking effect, which should reasonably affect all the stepwise procedures.

In their step-down procedure, Loughin and Noble consider $p - 2$ linear models, where at each step one estimate is constrained to zero (the related effect *is* in the model, but its estimate is zero). Doing so, they keep the residual deviance of the nested models equal to zero and let the total deviance $SST(\tilde{\mathbf{y}}_s) = SSE(\tilde{\mathbf{y}}_s)$ decrease at each step s of their algorithm.

Another possible choice is to keep the total deviance fixed and allow the residual deviance to increase through the steps of the algorithm. This can be done by always conditioning to the observed response and by considering a collection of nested linear models (where at each step the previously tested effects are removed from the model). Since the estimates of the effects are uncorrelated, the activeness of one effect can be evaluated through the amount of residual deviance due to removing that effect from the model. These are the basic ideas behind the tests we are going to introduce in this chapter. This section is dedicated to the definition of a suitable test statistic for 2^K unreplicated full factorial designs, whose distribution is provided in case errors are normally distributed and under a specific null hypothesis. Section 7.4 is dedicated to its permutation version, allowing control of the IER. A step-up procedure controlling IER or EER will be discussed in Section 7.5.

Let us briefly go back to the theory of linear models. Let the model (7.1)

$$Y_i = \beta_0 + \beta_1 x_{i1} + \beta_2 x_{i2} + \cdots + \beta_k x_{ik} + \cdots + \beta_{p-1} x_{i,p-1} + \varepsilon_i$$

be the complete model (possibly the saturated model as in unreplicated 2^K designs). Suppose we want to assess the null hypothesis $H_{0\beta_k} : \beta_k = 0$ against the alternative hypothesis $H_{1\beta_k} : \beta_k \neq 0$, $k = 1, \ldots, p-1$. Then $H_{0\beta_k}$ implies that data were generated by the reduced model

$$Y_i = \beta_0 + \beta_1 x_{i1} + \cdots + \beta_{k-1} x_{i,k-1} + \beta_{k+1} x_{i,k+1} + \cdots + \beta_{p-1} x_{i,p-1} + \varepsilon_i,$$

where the effect under testing β_k was removed (i.e., it was set equal to zero directly in the model). If there are degrees of freedom left to estimate the error variance (i.e., if $n > p$) and if we can assume normality of the error components, a way to test for $H_{0\beta_k}$ is to apply the F-test statistic, which is a ratio between two independent χ^2 random variables, namely

$$F = (n-p) \frac{SSR_{p-1} - SSR_p}{SSR_p}, \tag{7.5}$$

where SSR_p is the residual deviance of the complete model (with p parameters) and SSR_{p-1} is the residual deviance of the reduced model considering $p-1$ parameters. It is known that the difference between SSR_p and SSR_{p-1} is orthogonal to SSR_p and, under the assumption that the error components are normal, these random variables follow a χ^2 distribution. Under $H_{0\beta_k}$, the test statistic follows an F distribution with $1, n-p$ degrees of freedom. It is well known that (7.5) is equivalent to

$$F = (n-p) \frac{SSE_p - SSE_{p-1}}{SSR_p}, \tag{7.6}$$

where SSE_p and SSE_{p-1} stand for the explained deviance of the complete and reduced models. We stress that this can only be applied when we can assume normality of errors and when $n > p$.

In unreplicated full factorial designs, however, $n = p$ and $SSR_p = 0$. We must therefore choose a test statistic that does not depend on SSR_p. To this end, we recall the decomposition of the total deviance of Section 7.1. Since the estimates of the effects ar uncorrelated, we may write

$$SSE_p = n \left[\sum_{j \neq k} \hat{\beta}_j^2 + \hat{\beta}_k^2 \right],$$

$$SSE_{p-1} = n \sum_{j \neq k} \hat{\beta}_j^2.$$

and $SSE_p - SSE_{p-1} = n\hat{\beta}_k^2$. Moreover, SSE_{p-1} is a sum of $p-2$ uncorrelated random variables, and it is orthogonal to $SSE_p - SSE_{p-1}$. This is useful because it allows us to build a test statistic similar to (7.6) using the explained deviances instead of the residual ones. This test statistic has the form

$$T_F = (p-2) \frac{SSE_p - SSE_{p-1}}{SSE_{p-1}}. \tag{7.7}$$

The meaning of this particular test statistic is easy to understand: If the increase in the explained deviance of the model, when the kth parameter is present, is large enough w.r.t. the explained deviance of the model without the kth parameter, then β_k plays a significant role in fitting the response. On the other hand, if the increase in the explained variance of the model with the kth parameter is modest compared with the explained variance of the model without β_k, then H_{0k} should not be rejected.

In 2^K unreplicated full factorial designs, the test statistic (7.7) becomes

$$T_F^k = (p-2)\frac{\hat{\beta}_k^2}{\sum_{\substack{j=1 \\ j \neq k}}^{p-1} \hat{\beta}_j^2}. \tag{7.8}$$

When testing for active effects in unreplicated 2^K factorials, our interest is in applying separate tests for each effect (i.e., to test $H_{0\beta_1} : \{\beta_1 = 0\}$ against $H_{1\beta_1} : \{\beta_1 \neq 0\}$, irrespective of whether $H_{0\beta_2} \cup H_{0\beta_3} \cup \ldots \cup H_{0\beta_{p-1}}$ are true or not, $H_{0\beta_2}$ irrespective of whether $H_{0\beta_1} \cup H_{0\beta_3} \cup \ldots \cup H_{0\beta_{p-1}}$ are true or not, and so on). For each test, the null and the alternative hypotheses usually determine a partition of parametric space Θ^k into two regions Θ_0^k, Θ_1^k in such a way that $\Theta_0^k \cap \Theta_1^k = \emptyset, \Theta_0^k \cup \Theta_1^k = \Theta^k$, where Θ_0^k identifies those values of the parameter that are under the null hypothesis and Θ_1^k those that are under the alternative hypothesis for β_k. Let us first consider the parametric distribution of T_F. If the error components follow a normal distribution, then

$$\hat{\beta}_k \sim N(\beta_k, \sigma^2/n), \qquad\qquad k = 1, \ldots, p-1,$$

Since the $\hat{\beta}_k$s are uncorrelated (hence here they are also independent because they are normally distributed), the test statistic (7.8) follows an $F_{1,p-2}$ only if its denominator is a central χ^2 random variable. This implies that true effects β_j of the factors must be zero for all the effects in the denominator (i.e., we have to assume that $\beta_j = 0$, $j \neq k$). In other words, the test statistic (7.8) is suitable to assess the set of hypotheses $H_0^G = \cap_j H_{0\beta_j}$ against the alternative $H_{1\beta_k} : \{\beta_k \neq 0\} \cap \{\beta_j = 0, j \neq k\}$.

In a parametric framework, we can only test the region of the parameter space $\Theta^{k'} = \Theta^k \cap \{\cap_{j \neq k} \Theta_0^j\}$ and in general $\Theta^{k'} \subset \Theta^k$ and $\bigcup_j \Theta^{j'} \subset \Theta$. This implies that we are unable to test for the entire space Θ^k. The same restriction holds in a nonparametric framework, even though the distribution of the error components is not specified. Actually, the assumption $\beta_j = 0\ \forall j$ is required in order for the elements of the response to be exchangeable. This means that we can perform tests only in $\Theta^{k'} \subset \Theta^k, \forall k$. Note that whatever hypothesis is true for β_k, if normality of errors can be assumed and $\beta_j = 0, j \neq k$, the distribution of test statistic (7.8) is exact, but whenever there are some active effects among those in the denominator, the parametric distribution of (7.8) is no longer an F distribution because the denominator is a noncentral χ^2 random variable.

Since unreplicated two-level factorial designs are usually applied in screening experiments where little or nothing is known a priori about the effects, it does not seem possible to obtain exact testing for all active effects with (7.8). On the other hand, $\hat{\beta}_j$ is an unbiased estimate of $\beta_j, j = 1, \ldots, p-1$, and if $\hat{\beta}_k \geq \hat{\beta}_i$, then $T_F^k \geq T_F^i$, so the power of T_F monotonically follows the sizes of the effects. This happens because $SSE(p) = n\left[\hat{\beta}_k^2 + \sum_{j\neq k}\hat{\beta}_j^2\right]$. Hence the bigger $\hat{\beta}_k^2$ is, the smaller $SSE(p-1) = n\sum_{j\neq k}\hat{\beta}_j^2$ is.

7.4 The (Basso and Salmaso) T_P Test

The construction of a permutation test is based on the important notion of exchangeability of data. If this condition cannot be assumed, the test statistic is not exact (though it may still be unbiased and consistent).

In a permutation framework, the exchangeability of the elements of the response is usually determined by the null hypothesis. For instance, the response elements are independent (by assumption), and they are also identically distributed under the global null hypothesis H_0^G; therefore, under H_0^G, they are also exchangeable. That is, $\Pr(\mathbf{y}) = \Pr(\mathbf{y}^*)$, where $\mathbf{y}^*$ is any permutation of $\mathbf{y}$. Now let $\hat{\boldsymbol{\beta}}^* = [\hat{\beta}_1^*, \hat{\beta}_2^*, \ldots, \hat{\beta}_{p-1}^*]'$ be the vector of OLS estimates from $\mathbf{y}^*$ and $\mathbf{X}$. Then, under H_0^G, the $\hat{\beta}_j^*$'s are exchangeable, too, since they are independent linear combinations of exchangeable random variables. Indeed,

$$\Pr[\hat{\boldsymbol{\beta}}] = \Pr[(\mathbf{X}'\mathbf{X})^{-1}\mathbf{X}'\mathbf{y}] = \Pr[(\mathbf{X}'\mathbf{X})^{-1}\mathbf{X}'\mathbf{y}^*] = \Pr[\hat{\boldsymbol{\beta}}^*],$$

since $\mathbf{X}$ is a matrix of constants.

The same considerations can be done with respect to the vector of test statistics in (7.8), $\mathbf{T_F} = [T_F^1, T_F^2, \ldots, T_F^{p-1}]'$, which is function of $\hat{\boldsymbol{\beta}}$. Let $\mathbf{T_F^*}$ be the vector of test statistic (7.8) computed from $\mathbf{y}^*$. Then, under H_0^G, $\Pr[\mathbf{T_F}] = \Pr[\mathbf{T_F^*}]$.

Therefore, a permutation test for all $p-1$ effects can be provided by computing the reference distribution of the test statistic (7.8) for every possible permutation of the response $\mathbf{y}^*$ and every effect in the design. This implies that the test statistic for each effect T_F^k allows us to exactly test the global null hypothesis H_0^G against the single alternative $H_{1\beta_k} : \{\beta_k \neq 0\} \cap \{\beta_j = 0, j \neq k\}$. Since the global null hypothesis is required by all permutation tests on each effect, the null distribution of T_F^k can be obtained either by the permutation distribution of T_F^{k*} or the permutation distribution of the whole vector $\mathbf{T_F^*}$. Either choice leads to the same null distribution, so we choose to obtain a reference distribution for T_F^k by computing T_F^{k*} for any random permutation $\mathbf{y}^*$. As a final consideration, the scale factor $(p-2)$ in (7.8) can be omitted because the related distribution is permutationally invariant.

The previous consideration led us to define the permutation version of the T_F test, the T_P test, whose algorithm is as follows:

1. Obtain the estimates of observed effects $\hat{\beta}_j$, $j = 1, \ldots, p - 1$ from $\mathbf{y} = [y_1, \ldots, y_n]'$ and from design matrix $\mathbf{X}$. Let

$$T_P^k = \frac{\hat{\beta}_k^2}{\sum_{j \neq k}^{p-1} \hat{\beta}_j^2}, \qquad k = 1, \ldots, p - 1,$$

be the observed values of the test statistic for the effects $\beta_1, \beta_2, \ldots, \beta_{p-1}$.

2. Consider a large number B of permutations of $\mathbf{y}$ (in general, $B < n!$), and for each one compute

$$T_P^{k*} = \frac{\hat{\beta}_k^{*2}}{\sum_{j \neq k}^{p-1} \hat{\beta}_j^{*2}}, \qquad k = 1, \ldots, p - 1.$$

3. Compute p_k, the p-value related to the kth effect, as

$$p_k = \frac{\#[T_P^{k*} \geq T_P^k]}{B}.$$

The T_P test is exact for each effect under H_0^G, and hence it allows us to control the IER (see Table 7.1). However, the power of the test decreases as the number of active effects increases. This is a reasonable behavior that has also been observed in Loughin and Noble's test. In order to make a fair comparison with their test, we propose a simulation study along the lines of Loughin and Noble's (1997) paper. In their paper, they considered a growing number of active effects of the same size, and they reported the average rejection rates of active effects for each set of active effects. Three error distributions have been considered: normal, exponential, and t_3. All error variances were fixed at $\sigma^2 = 1$. Being a permutation test, its performance changed little across the three error distributions. This is also due to the fact that the estimates of effects are differences between two means of eight observations each, and evidently eight is a large enough number to make the central limit theorem take effect. Hence we decided to consider the same error distributions as in Loughin and Noble's paper, and we also considered the Cauchy distribution, which is a heavy-tailed distribution that is not suitable for the central limit theorem hypotheses (it does not have finite first and second moments). We compared the performances of T_P with those of Loughin and Noble's test across these four error distributions in a 2^4 design. Since T_P allows us to control the IER, we applied the calibration introduced in Loughin and Noble (1997) to control the same kind of error at an α level of 5%. The number of active effects goes from 1 to 13 in steps of 1, and we set their sizes equal to 0.5σ, σ, and 2σ. The results of the comparison are reported in Figure 7.1. Loughin and Noble's test (dotted line) is generally more powerful than T_P (solid line), as it can detect up to 11 active effects regardless of their size. T_P instead can only detect 7, 6 and 5 active effects when effects are equal to 0.5σ, σ, and 2σ, respectively. The results have been obtained for both tests by generating 1000 Monte Carlo data sets for each distribution considered and each scenario, both tests using $B = 1000$ permutations. There is a simple explanation for the behavior of T_P: The more active effects are present (and the greater they are), the bigger the noncentrality parameter in the denominator of the test statistic, reducing the potential for detecting significance. However, when few active effects are present, the power of T_P equals or exceeds that of Loughin and Noble's test.

The T_P procedure allows us to control the IER but there is no control of the EER. When each test is done at a level α, theoretically EER $\leq \alpha(p-1)$. The need for a conditional testing procedure to control the EER, together with considerations about T_P's rapid loss of power, suggest using a stepwise version of T_P, which is introduced in the next section.

Table 7.1. Observed IER of T_P on $2^K - 1 = 15$ effects under H_0^G with four different error distributions.

	Error Distributions			
α	$Norm$	Exp	t_3	Cau
0.01	0.011	0.009	0.010	0.012
0.05	0.049	0.050	0.052	0.051
0.10	0.101	0.101	0.101	0.099
0.15	0.150	0.149	0.149	0.150
0.20	0.202	0.202	0.193	0.199

7.5 The (Basso and Salmaso) Step-up T_P

The need for a stepwise procedure for T_P is motivated by the desire to control the EER and increase the T_P test power. The loss of power is due to the potential presence of active effects among those in the denominator of the test statistic. Thus, a step-down approach would not preserve the test from the *masking effect*. Furthermore, a step-down procedure should stop as soon as one effect is declared active because the null hypothesis on one effect implies that all the smaller effects are also inactive.

We therefore decided to test for the activeness of the effects according to their increasing order (given by the absolute-value ordering of the estimates of the observed effects), applying test statistic (7.8) in accordance with a sequence of nested null hypotheses that will be specified below. To do so, we assume the smallest effect $\beta_{(p-1)}$ is inactive. This assumption is necessary since the step-up procedure is basically a comparison among noncentralities of explained deviances of nested linear models, but we have nothing against which to compare the smallest effect. Along these lines, we assume that $H_{0\beta_{(p-1)}} : \beta_{(p-1)} = 0$ is true. This is reasonable since its corresponding estimate $\hat{\beta}_{(p-1)}$ is the smallest observed effect. We also recall that in Loughin

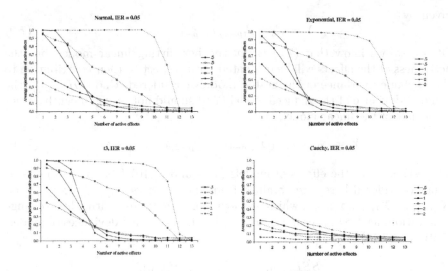

Fig. 7.1. Average rejection rates in a 2^4 unreplicated full factorial design with a growing number of active effects. T_p = solid line; Loughin and Noble's test = dotted line. Both tests control IER at 5%.

and Noble's permutation test we are allowed to test for all the effects except the smallest one.

Let $|\hat{\beta}_{(1)}| \geq |\hat{\beta}_{(2)}| \geq \cdots \geq |\hat{\beta}_{(p-2)}| \geq |\hat{\beta}_{(p-1)}|$ be the ordered absolute estimates of the effects. There are two main aspects to take into account: (1) the exchangeability of the response in accordance with a suitable null hypothesis; and (2) the noncentrality parameter in the denominator of (7.8), which should be zero in order to reduce the loss of power of T_P. The first aspect can be dealt with by considering a sequence of nested linear models to fit the observed response, each one corresponding to some "nested null hypotheses" that ensure the exchangeability of the response elements at each step of the procedure. The second aspect suggests that the testing procedure must stop whenever one effect is declared active by the testing procedure itself, whereas as long as no effects are declared active, the noncentrality parameter in the denominator of the test statistic can be assumed equal to zero.

The test statistic is still based on a ratio between explained variances as in (7.8), but the step-up procedure considers an increasing number of effects involved in the linear model. Given that $\beta_{(p-1)}$ is inactive, the step-up testing procedure starts with the test on $\beta_{(p-2)}$. The hypotheses to be assessed are

$$\begin{cases} H_{0\beta_{(p-2)}} : \beta_{(p-2)} = 0 \cap \{\beta_{(p-1)} = 0\} \\ H_{1\beta_{(p-2)}} : \beta_{(p-2)} \neq 0 \cap \{\beta_{(p-1)} = 0\} \end{cases}.$$

We have already pointed out that the response elements are not exchangeable. However, let us assume that the best-fitting linear model for the response is given by

$$Y_i = \beta_0 + \beta_{(p-1)}x_{i,(p-1)} + \varepsilon_i. \tag{7.9}$$

Of course, we know that (7.9) is *not* the best-fitting linear model, but the activeness of the effects will be evaluated with respect to their contribution to the explained variance of the nested models considered. Under $H_{0\beta_{(p-2)}}$, the residual deviance of the reduced model (7.9) should not be significantly larger than that of the linear model,

$$Y_i = \beta_0 + \beta_{(p-1)}x_{i,(p-1)} + \beta_{(p-2)}x_{i,(p-2)} + \varepsilon_i. \tag{7.10}$$

The estimates of the effects of models (7.9) and (7.10) (the overall mean $\hat{\beta}_0$ is not considered here) are obtained from the matrices $\mathbf{X}_{(p-1)} = \mathbf{x}_{(p-1)}$ and $\mathbf{X}_{(p-2)} = [\mathbf{X}_{(p-1)}, \mathbf{x}_{(p-2)}]$, which are orthogonal matrices. With a little change in notation, let $SSE_{(p-1)}$ and $SSE_{(p-2)}$ be the explained deviances of models (7.9) and (7.10). We have

$$SSE_{(p-2)} = n[\hat{\beta}^2_{(p-1)} + \hat{\beta}^2_{(p-2)}],$$
$$SSE_{(p-1)} = n[\hat{\beta}^2_{(p-1)}].$$

The observed value of the test statistic to assess $H_{0\beta_{(p-2)}}$ is then

$$T_P^{(p-2)} = 1 \cdot \frac{\hat{\beta}^2_{(p-2)}}{\hat{\beta}^2_{(p-1)}} = \frac{\mathbf{y}'[\mathbf{x}_{(p-2)}\mathbf{x}'_{(p-2)}]\mathbf{y}}{\mathbf{y}'[\mathbf{X}_{(p-2)}\mathbf{X}'_{(p-2)} - \mathbf{x}_{(p-2)}\mathbf{x}'_{(p-2)}]\mathbf{y}}, \tag{7.11}$$

which is the analogue of (7.8) in this context. In order to obtain the permutation distribution, we compute (7.11) for any random permutation of $\mathbf{y}$. That is, we compute

$$T_P^{(p-2)*} = 1 \cdot \frac{\hat{\beta}^{*2}_{(p-2)}}{\hat{\beta}^{*2}_{(p-1)}} = \frac{\mathbf{y}^{*'}[\mathbf{x}_{(p-2)}\mathbf{x}'_{(p-2)}]\mathbf{y}^*}{\mathbf{y}^{*'}[\mathbf{X}_{(p-2)}\mathbf{X}'_{(p-2)} - \mathbf{x}_{(p-2)}\mathbf{x}'_{(p-2)}]\mathbf{y}^*}.$$

Note that, doing so, the explained deviance of model (7.10) is partitioned in the same number of random variables either for the observed estimates or for the permutation estimates, and no corrections on the test statistic are needed. The p-value to test for $H_{0\beta_{(p-2)}}$ can therefore be calculated by computing

$$p_{(p-2)} = \frac{\#[T_P^{(p-2)*} \geq T_P^{(p-2)}]}{B}.$$

If $H_{0\beta_{(p-2)}}$ is rejected, then $\beta_{(p-2)}$ and all the greater effects $\beta_{(j)}, j = p - 3, \ldots, 1$ are declared active since $|\hat{\beta}_{(j)}| \geq |\hat{\beta}_{(p-2)}|$ for $j < p - 2$. On the other hand, if $H_{0\beta_{(p-2)}}$ is accepted, there is no evidence on observed data to consider $\beta_{(p-2)}$ active, and we can go on testing for the third smallest effect, $\beta_{(p-3)}$. Given that our testing procedure has recognized $\beta_{(p-2)}$ as inactive, the system of hypotheses on $\beta_{(p-3)}$ is

$$\begin{cases} H_{0\beta_{(p-3)}} : \beta_{(p-3)} = 0 \cap \{\beta_{(p-2)} = 0, \beta_{(p-1)} = 0\} \\ H_{1\beta_{(p-3)}} : \beta_{(p-3)} \neq 0 \cap \{\beta_{(p-2)} = 0, \beta_{(p-1)} = 0\} \end{cases}.$$

At this stage, we wish to evaluate whether the explained deviance of the model significantly increases by considering the complete model

$$Y_i = \beta_0 + \sum_{j=p-3}^{p-1} \beta_{(j)} x_{i(j)} + \varepsilon_i, \qquad i = 1, \ldots, n, \qquad (7.12)$$

instead of the reduced model

$$Y_i = \beta_0 + \sum_{j=p-2}^{p-1} \beta_{(j)} x_{i(j)} + \varepsilon_i, \qquad i = 1, \ldots, n. \qquad (7.13)$$

Now the OLS estimates of model (7.12) are obtained from $\mathbf{y}$ and the design matrix $\mathbf{X}_{(p-3)} = [\mathbf{X}_{(p-2)}, \mathbf{x}_{(p-3)}]$. The decomposition of the explained deviance is

$$SSE_{(p-3)} = n \left[\sum_{j=p-3}^{p-1} \hat{\beta}_{(j)}^2 \right],$$

$$SSE_{(p-2)} = n \left[\sum_{j=p-2}^{p-1} \hat{\beta}_{(j)}^2 \right].$$

The test statistic for $H_{\beta_{(p-3)}}$ becomes

$$\begin{aligned} T_P^{(p-3)} &= 2 \cdot \frac{\hat{\beta}_{(p-3)}^2}{\sum_{j=p-2}^{p-1} \hat{\beta}_{(j)}^2} \qquad (7.14) \\ &= 2 \cdot \frac{\mathbf{y}'[\mathbf{x}_{(p-3)} \mathbf{x}'_{(p-3)}]\mathbf{y}}{\mathbf{y}'[\mathbf{X}_{(p-3)} \mathbf{X}'_{(p-3)} - \mathbf{x}_{(p-3)} \mathbf{x}'_{(p-3)}]\mathbf{y}}. \end{aligned}$$

Again, the estimates of the effects obtained from the columns of $\mathbf{X}_{(p-3)}$ and $\mathbf{y}^*$ are exchangeable under $H_{0\beta_{(p-3)}}$, so the p-value for $\beta_{(p-3)}$ can be calculated by comparing the observed test statistic (7.14) with

$$\begin{aligned} T_P^{(p-3)*} &= 2 \cdot \frac{\hat{\beta}_{(p-3)}^{*2}}{\sum_{j=p-2}^{p-1} \hat{\beta}_{(j)}^{*2}} \\ &= 2 \cdot \frac{\mathbf{y}^{*'}[\mathbf{x}_{(p-3)} \mathbf{x}'_{(p-3)}]\mathbf{y}^*}{\mathbf{y}^{*'}[\mathbf{X}_{(p-3)} \mathbf{X}'_{(p-3)} - \mathbf{x}_{(p-3)} \mathbf{x}'_{(p-3)}]\mathbf{y}^*}, \end{aligned}$$

for each random permutation $\mathbf{y}^*$. The p-value for $\beta_{(p-3)}$ is then obtained as

$$p_{(p-3)} = \frac{\#[T_P^{(p-3)*} \geq T_P^{(p-3)}]}{B}.$$

If $H_{0\beta_{(p-3)}}$ is rejected, then $\beta_{(j)} \neq 0, j \geq p-3$, and we stop. Otherwise, we can go on to test for the fourth smallest effect $\beta_{(p-4)}$ and so on in a similar manner.

Note that this way of proceeding allows us to control aspects (1) and (2) mentioned at the beginning of this paragraph since the exchangeability of data is guaranteed by each null hypothesis, and condition (2) holds as long as there are no active effects among those in the denominator of the test statistic. We can generalize the step-up T_p procedure as follows:

1. Obtain the observed estimates of the effects from $\mathbf{X}$ and $\mathbf{y}$.
2. Order the observed estimates so that

$$|\hat{\beta}_{(p-1)}| \leq |\hat{\beta}_{(p-2)}| \leq \cdots \leq |\hat{\beta}_{(2)}| \leq |\hat{\beta}_{(1)}|.$$

3. At step $s = 2, 3, \ldots, 2^K - 1$, repeat:
   (a) Let $\mathbf{X}_{(p-s)} = [\mathbf{x}_{(p-1)}, \mathbf{x}_{(p-2)}, \ldots, \mathbf{x}_{(p-s)}]$.
   (b) Obtain the observed value of the test statistic:

$$T_P^{(p-s)} = (s-1) \cdot \frac{\mathbf{y}'[\mathbf{x}_{(p-s)}\mathbf{x}'_{(p-s)}]\mathbf{y}}{\mathbf{y}'[\mathbf{X}_{(p-s)}\mathbf{X}'_{(p-s)} - \mathbf{x}_{(p-s)}\mathbf{x}'_{(p-s)}]\mathbf{y}}. \quad (7.15)$$

   (c) Obtain the permutation distribution of (7.15) by computing B times the statistic

$$T_P^{(p-s)*} = (s-1) \cdot \frac{\mathbf{y}^{*'}[\mathbf{x}_{(p-s)}\mathbf{x}'_{(p-s)}]\mathbf{y}^*}{\mathbf{y}^{*'}[\mathbf{X}_{(p-s)}\mathbf{X}'_{(p-s)} - \mathbf{x}_{(p-s)}\mathbf{x}'_{(p-s)}]\mathbf{y}^*},$$

   where $\mathbf{y}^*$ is one of the B random permutations of $\mathbf{y}$.
   (d) Obtain the p-value of the test on $\beta_{(p-s)}$ by computing

$$p_{(p-s)} = \frac{\#[T_P^{(p-s)*} \geq T_P^{(p-s)}]}{B}$$

   (e) Finally, if $p_{(p-s)} \geq p_\alpha^{(p-s)}$, accept $H_{0\beta_{(p-s)}}$ and go back to point (a) with $s = s+1$; otherwise, declare the greater effects $\{\beta_{(j)}, j \leq p-s\}$ active and stop.

As long as there are no active effects in the reduced model (i.e., as long as $H_{0\beta_{p-s-1}}$ has not been rejected), the denominator of (7.15) acts as an estimate of the error variance. At step s of the algorithm, we have (unconditionally)

$$E[SSE_{(p-s+1)}|H_{0\beta_{(p-s+1)}}] = \frac{n}{s-1} \sum_{j=p-s+1}^{p-1} E[\hat{\beta}_{(j)}^2|H_{0\beta_{(p-s+1)}}] = \sigma^2,$$

$$E[n\hat{\beta}_{(p-s)}^2|H_{0\beta_{(p-s+1)}}] = \sigma^2 + n\beta_{(p-s)}^2.$$

The unconditional expected value of $T_p^{(p-s)}$ is

$$E[T_p^{(p-s)}|H_{0\beta_{(p-s+1)}}] = \frac{\sigma^2 + n\beta_{(p-s)}^2}{\sigma^2} = 1 + n\frac{\beta_{(p-s)}^2}{\sigma^2},$$

and the T_p statistic tends to assume high values whenever $H_{0\beta_{(p-s)}}$ is false.

At point (3e) of the algorithm, there is a stopping rule depending on a critical p-value $p_\alpha^{(p-s)}$. The calibration is needed since the step-up T_P is not an exact test. Moreover, being a sequential procedure, the probability of rejecting the null hypothesis at step s depends on what decision was made in the earlier steps. The critical p-values $p_\alpha^{(p-s)}$ also allow us to control the desired kind of error, IER or EER.

A possible way to control the IER is to perform the test at each step of the algorithm at the same significance level α. Note that the stopping rule should not be applied if IER is to be controlled. Under H_0^G, the estimate of the kth effect has probability $1/(p-1)$ of being the minimum (and hence not to be tested by the procedure) and probability $(p-2)/(p-1)$ of being tested. If no stopping rule is applied to the algorithm, and if the conditional probability of rejecting $H_{0\beta_{p-s}}$ is constantly equal to α, then the kth effect can be tested in any position, and the IER is

$$\begin{aligned}
\text{IER} &= \Pr[\hat{\beta}_k > \hat{\beta}_{(p-1)}]\Pr[\hat{\beta}_k \in \Theta_1^k|\hat{\beta}_k > \hat{\beta}_{(p-1)}] \\
&= \frac{p-2}{p-1}\sum_{s=2}^{p-1}\Pr[p_{(p-s)} \le p_\alpha^{(p-s)}|\hat{\beta}_k = \hat{\beta}_{(p-s)}] \cdot \Pr[\hat{\beta}_k = \hat{\beta}_{(p-s)}] \\
&= \frac{p-2}{p-1}\sum_{s=2}^{p-1}\alpha \cdot \frac{1}{p-2} \\
&= \frac{p-2}{p-1}\alpha.
\end{aligned}$$

Hence, by letting $\alpha = (p-1)/(p-2)\text{IER}$ we were able to control the IER. However, even though it is theoretically correct, this calibration is hard to justify since the stopping rule would not apply here (as in Loughin and Noble's step-down procedure). This is the reason why we recommend applying the T_P test of Section 7.4 in order to control the IER. However, a comparison between T_P and the step-up T_P calibrated to control the IER should be investigated.

We now investigate a possible calibration for EER, which is the probability of making a type I error on at least one effect when H_0^G is true. This probability depends on the number of tests that are to be made and the testing order. As mentioned before, the step-up procedure allows us to test up to $p-2$ effects. Note that the step-up T_P stops as soon as one effect is declared active. Thus, if H_0^G is true, a type I error may occur in any position. If α is the probability of making a type I error at each step of the step-up T_P algorithm, then

$$\text{EER} = 1 - (1-\alpha)^{p-2},$$

since EER $= 1 - \pi_0$, where π_0 is the probability of accepting H_0 for all the effects being tested. As a result, if we wish to control EER, we have to perform each single test with a significance level equal to

$$\alpha = 1 - (1 - \text{EER})^{\frac{1}{p-2}}.$$

The proof of this statement is as follows. If the probability of incorrectly declaring one effect active is held constant at each step of the algorithm, then, due to the step-up structure, the probability of incorrectly declaring the $(p - s)$th effect active when H_0^G is true becomes

$$\Pr[T_P^{(p-s)} \notin \Theta_0^{(p-s)} | H_0^G] = \alpha(1-\alpha)^{s-2} \qquad\qquad s = 2, 3, \ldots, 2^K - 1$$

since we have not rejected the null hypothesis $s - 1$ times before rejecting the null hypothesis on the $(p - s)$th effect. Hence:

$$1 - \text{EER} = Pr\left[\bigcap_{s=2}^{p-1} \left\{\beta_{(p-s)} \in \Theta_0^{(p-s)}\right\} | H_0^G\right]$$

$$= 1 - Pr\left[\bigcup_{s=2}^{p-1} \left\{\beta_{(p-s)} \notin \Theta_0^{(p-s)}\right\} | H_0^G\right]$$

$$= 1 - \alpha \sum_{s=2}^{p-1} (1-\alpha)^{s-2}$$

$$= 1 - \alpha \sum_{j=0}^{p-3} (1-\alpha)^{j}$$

$$= 1 - (1-\alpha)^{p-2}$$

since the sum in the fourth line is a geometric series with a common ratio $r = (1 - \alpha)$. Note that if $p - 2$ independent tests are made at a singificant level α, then the EER is the probability that the minimum p-value (among $p - 2$) is significant (that is, $\Pr[\min_s p_{(p-s)} \leq \alpha] = 1 - (1-\alpha)^{p-2}$), provided that each p-value is uniformly distributed under the null hypothesis. However, the sequential tests of the step-up T_P are *not* independent (and not even uniformly distributed), and therefore a Bonferroni correction applied directly to the significance level of each test would lead to a loss of power.

Either IER or EER calibration requires that the tests performed at each step of the algorithm reject the related null hypothesis with constant probability α. To do so, we must find some suitable critical p-values directly from the distribution of $p_{(p-s)}$. Some proposals for calibrating the step-up T_P test are considered in the next section.

7.5.1 Calibrating the Step-up T_p

The T_P test is a sequential algorithm comparing the explained variances of a reduced model and a complete model. It allows us to test for $p - 2$ effects over $p - 1$. The assumption of the T_P test is that the smallest estimated effect corresponds to a negligible effect.

There are basically three possible approaches to obtain a proper calibration, and all require the central limit theorem (CLT) to take effect. In fact, the OLS estimate of the jth effect can be written as

$$\hat{\beta}_j = \frac{1}{n}\mathbf{X}'_j\mathbf{y} = \frac{1}{n}\left[\sum_{x_{ij}=1} y_i + \sum_{x_{ij}=-1} y_i \right] = \frac{\bar{x}^+}{2} + \frac{\bar{x}^-}{2},$$

where $\bar{x}^+$ and $\bar{x}^-$ are sampling means of $n/2$ observations corresponding respectively to the $+1$ and -1 elements in the jth column of the design matrix. Since the response elements are independent random variables with variance $\sigma^2 < +\infty$ and if H_0^G holds, we have

$$E[\hat{\beta}_j] = 0,$$

$$V[\hat{\beta}_j] = \frac{\sigma^2}{n},$$

provided that $E[Y_i]$ and $V[Y_i]$ exist. Then, under H_0^G, as n increases

$$\sqrt{n}\frac{\hat{\beta}_j}{\sigma} \xrightarrow{d} N(0,1)$$

or, equivalently, $\hat{\beta}_j \xrightarrow{d} N(0, \sigma^2/n)$.

By recalling the CLT, we could provide a calibration for the desired EER (and IER as well; see the previous section) by considering a large number of independent data generations, by storing the distribution of $p_{(p-s)}$, and by choosing as the critical p-value for the sth step of the procedure a suitable α-quantile from that distribution. The choice of the quantile depends on which kind of error rate is required. We proved at the end of Section 7.5, that a calibration for EER can be obtained by either controlling the type I error at each step of the algorithm (which requires a critical p-value for each step) or by looking at the distribution of the minimum p-value.

How can we obtain the distribution of the p-values of the step-up T_P? We have considered three possible methods, which we will call "empirical", "Bonferroni" and "simulation".

The first choice is the "empirical" one: We may run the step-up T_P procedure a very large number of times with simulated standard normal data and no active effects, and each time store the p-values $\hat{p}_{(p-2)}, \hat{p}_{(p-1)}, \dots, \hat{p}_{(1)}$ produced by the step-up T_P. However, this way of proceeding is either approximate or time-expensive since it requires a lot of independent data generations to provide a good approximation of the distribution of $p_{(p-s)}$. The empirical critical p-values of the distribution of the $\min_s p_{(p-2)}$ for $K = 3, 4$ are reported in Table 7.2 in the "Emp" column. These calibrations are based on 1000 independent data generations with 1000 permutations each, and they are only reported here to give an idea of the sizes of the critical p-values. They would actually require a lot more permutations per each data generation since the minimum achievable significance level is related to the number of permutations considered (the empirical critical p-values should always be smaller than those obtained by a Bonferroni correction). This way of obtaining a calibration is not recommended unless one has an extremely powerful machine.

Another possible choice is the "Bonferroni" one: We may treat the sequential tests for each factor as independent and apply the Bonferroni correction to $p-2$ tests, which guarantees an upper bound for EER. This is equivalent to considering the limiting distribution of the test statistic at each step, as they were all independent. We know by the CLT that the test statistic at step s, as $n \to +\infty$, satisfies

$$T_P^{(p-s)} = (s-1) \frac{n \hat{\beta}_{(p-s)}^2 / \sigma^2}{n \sum_{j=p-s+1}^{p-1} \hat{\beta}_{(j)}^2 / \sigma^2} \xrightarrow{d} F_{1,s-1} \qquad s = 2, 3, \ldots, 2^K - 1.$$

We could then directly simulate the distribution of the sequential test statistics. Equivalently, we could set the critical p-values equal to $p_{\alpha*}^{(p-2)} = 1 - (1 - EER)^{1/(p-2)}$ if EER is to be controlled. The critical p-values of this procedure are reported in the "Bonferroni" column of Table 7.2. This choice will save time but will also be a bad approximation of the real critical values (or p-values) since the sequential tests are treated as independent, and they are not.

The "simulation" choice is another possible way, and it seems to be the best one because it is quick and accurate. Since it does not seem easy to obtain the dependence structure of the sequential test statistics, the better way to obtain a calibration is probably just to simulate the entire process:

- Consider a large number of independent data generations, say G. For $g \in 1, \ldots, G$ repeat:
  1. Obtain $p-1$ random observations from $N(0, n^{-1})$. These observations will simulate the estimates of the effects ${}^g\hat{\beta}_1, {}^g\hat{\beta}_2, \ldots, {}^g\hat{\beta}_{p-1}$.
  2. Order the simulated estimates of the effects $|{}^g\hat{\beta}_{(p-2)}| \leq |{}^g\hat{\beta}_{(p-1)}| \leq \cdots \leq |{}^g\hat{\beta}_{(1)}|$.
  3. Obtain the simulated sequential test statistics

$$ {}^g T_P^{(p-s)} = (s-1) \frac{{}^g\hat{\beta}_{(p-s)}^2}{\sum_{j=p-s+1}^{p-1} {}^g\hat{\beta}_{(j)}^2}, \qquad s = 2, 3, \ldots, 2^K - 1.$$

  4. Obtain the estimate of the sequential p-values as

$$ {}^g p^{(p-2)} = 1 - F_\Phi({}^g T_P^{(p-s)}), \qquad \text{where} \quad \Phi \sim F_{1,s-1},$$

$$ s = 2, 3, \ldots, 2^K - 1.$$

At this point, let $\mathbf{P}$ be a $G \times (p - 2)$ matrix, where the p-values related to each data generation have been stored according to the preferred calibrating choice (the "empirical", or "simulation" one). Note that the first column of $\mathbf{P}$ contains the simulated p-values for the test on $\beta_{(p-2)}$, the second column contains the simulated p-values for the test on $\beta_{(p-3)}$, and so on.

In order to obtain a calibration for IER, the critical p-values for each step of the procedure are the IER-quantiles of the columns of $\mathbf{P}$. Note that if one wants to control the IER with the step-up T_P, he should not apply the stopping rule.

Critical values to control the EER can be found by either choosing the $\tilde{\alpha}$-quantiles of the columns of $\mathbf{P}$, where $\tilde{\alpha} = 1 - (1 - \text{EER})^{1/(p-2)}$, or by choosing $p_\alpha^{(p-s)} = p_\alpha$ as the EER-quantile of the $^g\min_p$ distribution, where $^g\min_p$ is the minimum p-value of the gth row of $\mathbf{P}$, $g = 1, \ldots, G$.

The column "Sim" of Table 7.2 reports the critical p-values obtained by applying the "simulation" choice for $K = 3, 4$. Note that the critical values are always smaller than the Bonferroni ones (this is because the sequential tests are dependent), and they are very close to the empirical ones. Since obtaining the critical p-values at each step is equivalent, in order to control EER, to obtain a single critical value given by the $^g\min_p$ distribution, we have chosen this second way in the R function in Section 7.7.

Table 7.2. Empirical, Bonferroni and simulated critical p-values of the T_p test.

Critical p-values, $K = 3$			
EER	Emp	Bonf	Sim
0.01	0.00100	0.00851	0.00062
0.05	0.00494	0.01741	0.00328
0.1	0.00799	0.02672	0.00669
0.2	0.01598	0.03651	0.01320

Critical p-values, $K = 4$			
EER	Emp	Bonf	Sim
0.01	0.00047	0.00366	0.00042
0.05	0.00188	0.00750	0.00193
0.1	0.00357	0.01154	0.00364
0.2	0.00698	0.01581	0.00728

7.6 A Comparative Simulation Study

In this section we propose a comparative simulation study between the step-up T_P and Loughin and Noble's test controlling for EER. The chosen level of EER is 0.1, and the simulation study is similar to the one presented in

Section 7.4. Four error distributions are considered: normal, exponential, t_3, and Cauchy. A growing number of active effects, from 1 to 13, determines each scenario. The sizes of active effects are 0.5σ, σ, and 2σ, where σ^2 is set equal to 1. Results are reported in Figure 7.2. For each error distribution, the average rejection rates of the active effects are reported for both the step-up T_P (solid line) and Loughin and Noble's test (dotted line). The horizontal solid line (called "H0 Stp-Tp") indicates the step-up T_P-achieved EER when the global null hypothesis is true. The horizontal dotted line (called "H0 LNT") is the equivalent for Loughin and Noble's test. Any point beyond the horizontal lines means that the corresponding number of active effects has been recognized by the related test, at least regarding the average power. The results were obtained for both tests by generating 1000 Monte Carlo data sets for each distribution considered and each scenario, both tests applying $B = 1000$ permutations.

The behavior of both tests varies little among different error distributions as an effect of the central limit theorem. The power of Loughin and Noble's test decreases as the number of active effects increases, while the power of the step-up T_P seems to remain stable, although it does not show a monotone trend. This is due to two contrasting aspects. On the one hand, the more active effects present, the more difficult it is for the step-up T_P to recognize them from the noise. On the other hand, the step-up T_P is a conditional procedure, where the effect $\beta_{(p-s)}$ is tested only if $\beta_{(p-s+1)}$ has been tested. The greater s, the smaller the probability of being tested.

In order to show the good behavior of the step-up T_P, we consider the same example reported in Loughin and Noble (1997). This example is Problem 9.11 from Montgomery (1991) concerning a process used to make an alloy for jet engine components. It is a four-factor unreplicated two-level full factorial design, where two replicates are available. This also allows us a comparison with parametric ANOVA. The permutation tests will be run on the average response for each treatment. This example will also be investigated in Section 7.7. We refer both to Montgomery (1991) and to Loughin and Noble (1997) for further details.

Loughin and Noble showed that their test is able to detect nine active effects in the example considered using EER = 0.2. Table 7.3 shows the results of the comparison between Loughin and Noble's permutation test and the step-up T_P in this specific example. In the first column, the labels of the estimates and the OLS estimates of the effects are reported. The second and third columns refer to Loughin and Noble's test (LNT), and here the p-values they have obtained with this example and the corresponding decision rule when EER = 0.1 are reported (0 means inactive and 1 means active). In their work, the suggested calibrations for a 2^4 design are 0.042 for EER = 0.05, 0.075 for EER = 0.10, and 0.135 for EER = 0.20. Considering EER = 0.2, there are nine significant effects since the eighth p-value is smaller than the 0.2 critical value 0.135, which is reported in the the last row of the second

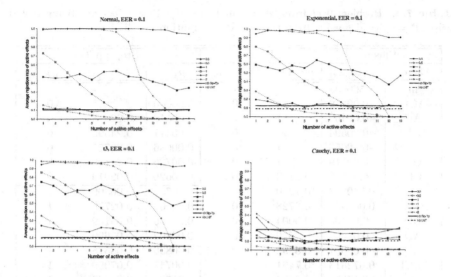

Fig. 7.2. Average rejection rates in a 2^4 unreplicated full factorial design with a growing number of active effects. T_p = solid line, Loughin and Noble's test = dotted line. Both tests control EER at 10%.

column. Instead, if we choose EER = 0.1, no effects are declared significant, since there are no p-values smaller than the 0.1 critical value, which is 0.075.

The last three columns refer to the step-up T_p. The first one reports the critical p-values $p_{\tilde{\alpha}}^{(p-2)}$ related to each step of the algorithm that were obtained through the "simulation" method described above. The critical p-values are the $\tilde{\alpha}$-quantiles of the distribution of ${}^g p_{(p-s)}$, where $\tilde{\alpha} = 1 - 0.9^{1/14}$. They can be interpreted either as the critical values for IER = 0.0075 or for EER = 0.1. In the first case, only the effect BC is declared active. In the latter case, there are nine significant effects since the procedure stops as soon as one effect is declared active, and it is still the case of factor BC. The last row of the sixth column reports the critical p-value obtained through the "simulation" method and the distribution of the minimum p-value. This approach leads to the same results since the only p-value smaller than 0.00362 is that of factor BC. Note that the effect of factor ACD has not been tested (the R program we implemented always returns $p_{(p-1)} = 1$). Nine effects are declared significant by our step-up T_P procedure with EER = 0.1. The step-up T_P was run with $B = 5000$.

We have also run the T_P test on the same example with IER = 0.05. The test was able to detect a single effect (that of factor C, the largest one). With the same IER, the step-up T_P with no stopping rule (controlling the IER) revealed as active the effects of factors CD and BC.

Table 7.3. Problem 9.11 from Montgomery (1991); 1^* = effect not being tested; critical* = critical p-value from the min-P distribution for EER = 0.1.

Effect		LNT		Step-up T_p		
Factor	$\hat{\beta}_{(p-s)}$	p-value	$D_{EER=0.1}$	$p_{\tilde{\alpha}}^{(p-2)}$	$p_{(p-s)}$	$D_{EER=0.1}$
m	1.560938	-	-		-	-
ACD	−0.00219	1	0	0.00000	1^*	0
AC	−0.00469	1	0	0.00510	0.27615	0
BCD	−0.00594	1	0	0.00411	0.24395	0
ABD	−0.00656	1	0	0.00636	0.23755	0
ABC	−0.00719	1	0	0.00726	0.23195	0
CD	−0.01719	0.3113	0	0.00626	0.02919	0
BC	0.040313	0.1260	0	0.00781	0.00300	1
AB	0.045938	0.3728	0	0.00967	0.02779	1
BD	−0.05406	0.5004	0	0.00936	0.05159	1
AD	−0.05656	0.7529	0	0.00892	0.07699	1
B	−0.06031	0.8298	0	0.00836	0.08578	1
ABCD	0.072813	0.7461	0	0.00741	0.07099	1
A	−0.10656	0.3094	0	0.00501	0.02240	1
D	0.112188	0.5087	0	0.00281	0.03439	1
C	−0.14594	0.3168	0	0.00048	0.02260	1
critical$*$		**0.075**			**0.00362**	

7.7 Examples with R

Let's begin with the design matrix. The function `create.design` creates a Hadamard matrix of order $n = 2^K$, where K is the entry of the function (factor labels are available up to $k = 6$, although the function can create designs of greater orders). For instance, let $K = 3$, and then type

```
> source("Path/create_design.r")
> k=3
> X<-create.design(k)
> X
     m  A  B  C AB AC BC ABC
[1,] 1  1  1  1  1  1  1   1
[2,] 1  1  1 -1  1 -1 -1  -1
[3,] 1  1 -1  1 -1  1 -1  -1
[4,] 1  1 -1 -1 -1 -1  1   1
[5,] 1 -1  1  1 -1 -1  1  -1
[6,] 1 -1  1 -1 -1  1 -1   1
[7,] 1 -1 -1  1  1 -1 -1   1
[8,] 1 -1 -1 -1  1  1  1  -1
```

Note that the design matrix is not in standard Yates order. If Yates order is required, type:

```
> X<-create.design(k,Yates=TRUE)
> X
     m  A  B  C AB AC BC ABC
[1,] 1 -1 -1 -1  1  1  1  -1
[2,] 1  1 -1 -1 -1 -1  1   1
[3,] 1 -1  1 -1 -1  1 -1   1
[4,] 1  1  1 -1  1 -1 -1  -1
[5,] 1 -1 -1  1  1 -1 -1   1
[6,] 1  1 -1  1 -1  1 -1  -1
[7,] 1 -1  1  1 -1 -1  1  -1
[8,] 1  1  1  1  1  1  1   1
```

Now let's generate the vector of data under the global null hypothesis $\beta_1 = \beta_2 = \cdots = \beta_7 = 0$:

```
>   set.seed(101)
>   n<-2^k
>   b<-rep(0,n)
>   e<-rnorm(n)
>   y<-X%*%b+e
>   y
              [,1]
[1,] -0.3260365
[2,]  0.5524619
[3,] -0.6749438
[4,]  0.2143595
[5,]  0.3107692
[6,]  1.1739663
[7,]  0.6187899
[8,] -0.1127343
```

Here b is the vector of the true effects $\beta_1, \beta_2, \ldots, \beta_7$. The function unreplicated allows us to perform the analysis of a 2^K unreplicated factorial design either by applying the T_P (controlling IER) or the step-up T_P test (controlling EER), at the specified type error rates. Note that the T_P test does not need calibration, whereas the step-up T_P does. The function requires as entries the design matrix X and the array of responses y. By default, the unreplicated.r function carries out the T_P test test controlling IER=0.05 and considering B = 1000 permutations. To analyze data, load the functions unreplicated.r and T_to_P.r (which computes the p-values) and run the analysis by typing:

```
> source("Path/unreplicated.r")
> source("Path/t2p.r")
> t<-unreplicated(y,X)
> t$beta
$beta
            [,1]
```

```
m     0.21957900
A     0.23743432
B    -0.20821121
C     0.27811876
AB   -0.19798954
AC   -0.20451609
BC   -0.03645878
ABC  -0.20069077
```

```
> table<-cbind(t(t$p.value),t(t$dec))
> colnames(table)<-c("p.value","Decision")
> table
      p.value Decision
A    0.3026973        0
B    0.3486513        0
C    0.1918082        0
AB   0.4255744        0
AC   0.3596404        0
BC   0.8931069        0
ABC  0.4055944        0
```

Note that the vector of estimates contains the intercept, whereas statistical tests are only made on factor effects.

The output of the unreplicated function is a list that displays the estimates of the effects (beta, where "m" stands for the overall mean, corresponding to the first column of **X**), the related p-values, and decisions. By decision equal to 1, we mean that the related effect is significant at an IER = 0.05. In order to manipulate the objects in the output, assign the function to an object U and then recall the objects in the list. For instance, in order to display a summary table of the analysis with an IER = 0.2, type:

```
> set.seed(101)
> U<-unreplicated(y,X,IER=0.2)
> table<-cbind(U$beta[-1],t(U$p.value),t(U$dec))
> colnames(table)=c("beta","p.value","Decision")
> table
            beta    p.value Decision
A     0.23743432 0.3026973        0
B    -0.20821121 0.3486513        0
C     0.27811876 0.1918082        1
AB   -0.19798954 0.4255744        0
AC   -0.20451609 0.3596404        0
BC   -0.03645878 0.8931069        0
ABC  -0.20069077 0.4055944        0
```

Note how the decision related to the main factor C is now equal to 1 since we have set the IER equal to 0.2. With the T_P test, all $2^K - 1$ main effects and interactions are tested.

Let's investigate what happens when there are two active effects and the IER is set equal to 0.05:

```
> b[2:3]<-1
> y<-X%*%b+rnorm(8)
> set.seed(101)
> U<-unreplicated(y,X)
> table<-cbind(U$beta[-1],t(U$p.value),t(U$dec))
> colnames(table)=c("beta","p.value","Decision")
> table
          beta     p.value Decision
A    0.1536359 0.80519481        0
B    1.0458663 0.02597403        1
C    0.3569429 0.55244755        0
AB  -0.1263682 0.85414585        0
AC  -0.3234737 0.62937063        0
BC  -0.3092982 0.69330669        0
ABC  0.5419082 0.35164835        0
```

In this case, only the largest estimated effect has been declared significant. This happens because when the noncentrality parameter of factor B (β_B) is in the numerator of T_P, the resulting test statistic assumes large values ($T_P^B = 1.6550$), whereas in the remaining cases β_B is in the denominator of the T_P statistic, so its contribution produces small values of the test statistic. To obtain the observed values of the test statistic for each effect, type:

```
> beta<-U$beta[-1]
> SSE<-sum(beta^2)
> T.oss<-beta^2/(SSE-beta^2)
> names(T.oss)<-colnames(X)[-1]
> round(T.oss,digits=4)
     A      B      C     AB     AC     BC    ABC
0.0136 1.6550 0.0783 0.0092 0.0634 0.0577 0.2010
```

That is the reason for the decreasing power of the T_P test as the number of active effects increases. To see what happens with a random permutation of the response, type:

```
> set.seed(10)
> y.star<-sample(y)
> beta.star<-unreplicated(y.star,X,3)$beta[-1]
> SSE.star<-sum(beta.star^2)
> T.star<-beta.star^2/(SSE.star-beta.star^2)
> names(T.star)<-colnames(X)[-1]
> round(T.star,digits=4)
```

```
        A       B       C      AB      AC      BC     ABC
    0.3757  0.0005  0.4955  0.2612  0.1348  0.0007  0.0736
```

Note how this permutation value of T_P^{B*} is smaller than the observed value T_P^B, whereas this does not happen for factor A (the estimated probability of the event $\{T_P^{A*} < T_P^A\} = 1 - p_A$ is equal to 0.194, and that of the event $\{T_P^{B*} < T_P^B\} = 1 - p_B$ is 0.974).

Let's now consider Problem 9.11 from Montgomery (1991). This is a 2^4 design matrix in Yates order, where two replicates per treatment are available. This allows us to run the parametric ANOVA test:

```
> d<-read.csv("Mont_9.11.csv",header=TRUE)
> y<-c(d[,1],d[,2])
> X<-create.design(4,Yates=TRUE)
> A<-factor(rep(X[,2],2))
> B<-factor(rep(X[,3],2))
> C<-factor(rep(X[,4],2))
> D<-factor(rep(X[,5],2))
> summary(aov(y~A*B*C*D))
```

	Df	Sum Sq	Mean Sq	F value	Pr(>F)	
A	1	0.36338	0.36338	54.7205	1.509e-06	***
B	1	0.11640	0.11640	17.5289	0.0006975	***
C	1	0.68153	0.68153	102.6301	2.292e-08	***
D	1	0.40275	0.40275	60.6499	7.846e-07	***
A:B	1	0.06753	0.06753	10.1689	0.0057096	**
A:C	1	0.00070	0.00070	0.1059	0.7490981	
B:C	1	0.05200	0.05200	7.8311	0.0128837	*
A:D	1	0.10238	0.10238	15.4169	0.0012048	**
B:D	1	0.09353	0.09353	14.0842	0.0017371	**
C:D	1	0.00945	0.00945	1.4235	0.2502161	
A:B:C	1	0.00165	0.00165	0.2489	0.6246116	
A:B:D	1	0.00138	0.00138	0.2075	0.6548322	
A:C:D	1	0.00015	0.00015	0.0231	0.8812025	
B:C:D	1	0.00113	0.00113	0.1699	0.6856858	
A:B:C:D	1	0.16965	0.16965	25.5478	0.0001173	***
Residuals	16	0.10625	0.00664			

```
---
Signif. codes:0 '***' 0.001 '**' 0.01 '*' 0.05 '.' 0.1' '1
```

The ANOVA detects nine effects as significant. Let's now run the step-up T_P controlling EER = 0.1. We will use as the response the average response for each treatment. To obtain the data, type:

```
> y = apply(d,1,mean)
> y
```

1	2	3	4	5	6	7	8	9	10
1.810	1.450	1.440	1.610	1.305	1.255	1.440	1.280	2.115	1.855

11	12	13	14	15	16
1.870	1.505	1.865	1.315	1.495	1.365

The step-up T_P requires an EER calibration. We could run the procedure with the critical p-values indicated in Table 7.2. At the end of this section, the code to obtain the "simulation" method of calibration is reported for a 2^4 design. Type:

```
> set.seed(10)
> X<-create.design(4,Yates=TRUE)
> U<-unreplicated(y,X,step.up=TRUE,EER=0.00364,B=5000)
> table<-cbind(U$beta[-1],t(U$p.value),t(U$dec))
> colnames(table)=c("beta","p.value","Decision")
> table
```

	beta	p.value	Decision
A	-0.1065625	0.02479504	1
B	-0.0603125	0.08818236	1
C	-0.1459375	0.02119576	1
D	0.1121875	0.03619276	1
AB	0.0459375	0.02939412	1
AC	-0.0046875	0.27374525	0
AD	-0.0565625	0.07918416	1
BC	0.0403125	0.00339932	1
BD	-0.0540625	0.04319136	1
CD	-0.0171875	0.03439312	0
ABC	-0.0071875	0.22955409	0
ABD	-0.0065625	0.23655269	0
ACD	-0.0021875	1.00000000	0
BCD	-0.0059375	0.24855029	0
ABCD	0.0728125	0.06798640	1

The step-up T_P controlling EER = 0.1 gives the same results as the ANOVA test. Now let's represent the Q-Q normal plot of the estimates of the effects by highlighting the active effects detected by the step-up T_P. Type:

```
> t<-qqnorm(table[,1],ylim=c(-0.2,0.2))
> qx<-t$x
> qy<-t$y
> dec<-table[,3]
> points(qx[dec==1],qy[dec==1],pch=20)
```

This will produce the output in Figure 7.3.

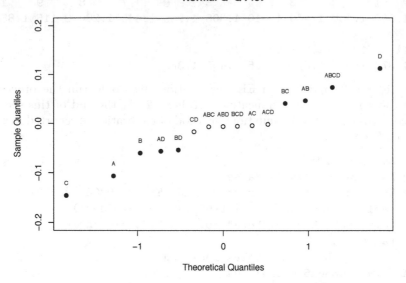

Fig. 7.3. Normal Q-Q plot of Problem 9.11 of Montgomery (1991).

7.7.1 Calibrating the Step-up T_P with R

The "simulation" method to obtain a calibration for EER in the step-up T_P was described in Subsection 7.5.1. The code to calibrate the step-up T_P at EER equal to .01, .05, .1, and .2 by simulating the step-up T_P procedure with $G = 10,000$ data generations is as follows:

```
> set.seed(101)
> k=4
> n=2^k
> G=10000
> F<-array(0,dim=c(G,n-1))
> P<-array(0,dim=c(G,n-1))
> for(cc in 1:G){
> b<-sort(abs(rnorm(n-1,sqrt(1/n))))
> for(j in 2:(n-1)){
> F[cc,j]<-(j-1)*b[j]^2/(sum(b[1:j]^2)-b[j]^2)
> }
> }
> for(j in 2:(n-1)){
> P[,j]<- 1-pf(F[,j],1,(j-1))
> }
> min<-apply(P[,-1],1,min)
> EER<-c(0.01,0.05,0.1,0.2)
```

```
> quantile(min,EER)
            1%              5%            10%            20%
0.0003631566 0.0018964540 0.0037944725 0.0073475314
```

The calibration only requires some seconds to be run. For better approximations, increase the number of data generations G. The results are the critical p-values obtained from the $^g\min_p$ distribution. The critical p-values for each step of the step-up T_P that guarantee an EER $= 0.1$ can be obtained by typing:

```
> p<-dim(X)[2]
> EER<-0.1
> tilde.alpha<-1-(1-EER)^(1/(p-2))
> critical<-apply(P,2,function(x){quantile(x,tilde.alpha)})
> critical<-data.frame(step=seq(1,15),critical=critical)
> critical
   step     critical
1     1 0.0000000000
2     2 0.0053802258
3     3 0.0050186952
4     4 0.0051173826
5     5 0.0064026723
6     6 0.0074743545
7     7 0.0093800891
8     8 0.0097069040
9     9 0.0096165022
10   10 0.0088048292
11   11 0.0089946047
12   12 0.0071454521
13   13 0.0046289031
14   14 0.0028677822
15   15 0.0004491715
```

References

ABELSON, R. B. and TUCKEY, J. W. (1963). Efficient utilization of non-numerical information in quantitative analysis: general theory and the case of simple order. *Annals of Mathematical Statistics* **34**, 1347–1369.

AGRESTI, A. (2002). *Categorical Data Analysis*. Wiley, Hoboken.

AGRESTI, A. and COULL, B. A. (2002). The analyis of contingency tables under inequality constraints. *Journal of Statistical Planning and Inference* **107**, 45–73.

AGRESTI, A. and KLINGENBERG, B. (2005). Multivariate tests comparing binomial probabilities, with application to safety studies for drugs. *Applied Statistics* **54**, 691–706.

AGRESTI, A. and LIU, I.-M. (1999). Modelling a categorical variable allowing arbitrarily many category choices. *Biometrics* **55**, 936–943.

ANDERSON, M. J. (2001). A new method for non-parametric multivariate analysis of variance. *Austral Ecology* **26**, 32–46.

ANDERSON, M. J. and TER BRAAK, C. J. F. (2003). Permutation tests for multi-factorial analysis of variance. *Journal of Statistical Computation and Simulation* **73**, 85–113.

ARMITAGE, P. (1955). Tests for linear trends in proportions and frequencies. *Biometrics* **11**, 375–386.

BAIRD, S. J., CATALANO, P. J., RYAN, L. M. and EVANS, J. S. (1997). Evaluation of effect profiles: functional observational battery outcomes. *Fundamental and Applied Toxicology* **40**, 37–51.

BASSO, D., CHIARANDINI, M. and SALMASO, L. (2007). Synchronized permutation tests in replicated I × J designs. *Journal of Statistical Planning and Inference* **137**, 25642578.

BASSO, D. and SALMASO, L. (2006). A discussion on permutation tests conditional to observed responses in unreplicated 2^M factorial designs. *Communications in Statistics: Theory and Methods* **35**, 83–97.

BASSO, D. and SALMASO, L. (2007). A comparative study of permutation tests for unreplicated two-level experiments. *Journal of Applied Statistical Science* **15**, 87–98.

BERGER, V. W. (1998). Admissibility of exact conditional tests of stochastic order. *Journal of Statistical Planning and Inference* **66**, 39–50.

BERGER, V. W. and IVANOVA, A. (2002a). Adaptive tests for ordered categorical data. *Journal of Modern Applied Statistical Methods* **1**, 269–280.

BERGER, V. W. and IVANOVA, A. (2002b). The bias of linear rank tests when testing for stochastic order in ordered categorical data. *Journal of Statistical Planning and Inference* **107**, 237–247.

BILDER, C. R., LOUGHIN, T. M. and NETTLETON, D. (2000). Multiple marginal independence testing for pick any/c variables. *Communications in Statistics: Simulation and Computation* **29**, 1285–1316.

BITTMAN, R. M., ROMANO, J. P., VALLARINO, C. and WOLF, M. (2006). Testing multiple hypotheses with common effect direction using the closure method. Working Paper .

BRUNNER, E. and PURI, M. (2001). Nonparametric methods in factorial designs. *Statistical Papers* **42**, 1–52.

CADE, B. S. and RICHARDS, J. D. (1996). Permutation tests for least absolute deviation regression. *Biometrics* **52**, 886–902.

CHUANG-STEIN, C. and AGRESTI, A. (1997). Tutorial in biostatistics: a review of tests for detecting a monotone dose-response relationship with ordinal response data. *Statistics in Medicine* **16**, 2599–2618.

COCHRAN, W. G. (1954). Some methods for strengthening the common χ^2 tests. *Biometrics* **10**, 417–451.

COHEN, A., MADIGAN, D. and SACKROWITZ, H. B. (2003). Effective directed tests for models with ordered categorical data. *Australian and New Zeland Journal of Statistics* **45**, 285–300.

COHEN, A. and SACKROWITZ, H. B. (1991). Tests for independence in contingency tables with ordered categories. *Journal of Multivariate Analysis* **36**, 57–67.

COHEN, A. and SACKROWITZ, H. B. (1998). Directional tests for one-sided alternatives in multivariate models. *The Annals of Statistics* **26**, 2321–2338.

COHEN, A. and SACKROWITZ, H. B. (2000). Testing whether treatment is "better" than control with ordered categorical data: definitions and complete class theorems. *Statistics and Decisions* **18**, 1–25.

COHEN, A. and SACKROWITZ, H. B. (2004). A discussion of some inference issues in order restricted models. *The Canadian Journal of Statistics* **32**, 199–205.

COHEN, A., SACKROWITZ, H. B. and SACKROWITZ, M. (2000). Testing whether treatment is "better" than control with ordered categorical data: an evaluation of new methodology. *Statistics in Medicine* **19**, 2699–2712.

COX, D. R. and HINKLEY, D. V. (1974). *Theoretical Statistics*. Chapman and Hall, London.

DANIEL, C. (1959). Use of half-normal plots in interpreting factorial two-level experiments. *Technometrics* **1**, 311–340.

DEAN, A. and VOSS, D. (1999). *Design and Analysis of Experiments*. Springer-Verlag, New York.

DEMING, W. E. and STEPHAN, E. F. (1940). On a least squares adjustment of a sampled frequency table when the expected margins are known. *Annals of Mathematical Statistics* **11**, 427–444.

DIETZ, E. J. (1998). Multivariate generalizations of Jonckheere's test for ordered alternatives. *Communications in Statistics: Theory and Methods* **18**, 3763–3783.

DMITRIENKO, A., OFFEN, W. W. and WESTFALL, P. H. (2003). Gatekeeping strategies for clinical trials that do not require all primary effects to be significant. *Statistics in Medicine* **22**, 2387–2400.

DMITRIENKO, A. and TAMHANE, A. C. (2007). Gatekeeping procedures with clinical trial applications. *Pharmaceutical Statistics* **6**, 171–180.

DRAPER, D. (1988). Rank-based robust analysis of linear models. i. exposition and review. *Technometrics* **3**, 239–271.

DYKSTRA, R. L., LEE, C. C. and YAN, X. (1996). Multinomial estimation for two stochastically ordered distributions. *Statistics and Probability Letters* **30**, 353–361.

EDGINGTON, E. S. (1995). *Randomization Tests*. Marcel Dekker, New York, 3rd ed.

ELBARMI, H. and MUKERJEE, H. (2005). Inferences under a stochastic ordering constraint: the k-sample case. *Journal of the American Statistical Association* **100**, 252–261.

GAIL, M. (1974). Value systems for comparing two independent multinomial trials. *Biometrika* **61**, 91–100.

GRAUBARD, B. I. and KORN, E. L. (1987). Choice of column scores for testing independence in ordered $2 \times k$ tables. *Biometrics* **43**, 471–476.

GROSS, S. T. (1981). On asymptotic power and efficiency of tests of independence in contingency tables with ordered classifications. *Journal of the American Statistical Association* **76**, 935–941.

HAMADA, M. and BALAKRISHNAN, N. (1998). Analyzing unreplicated factorial experiments: a review with some new proposals. *Statistica Sinica* **8**, 1–41.

HAN, K. E., CATALANO, P. J., SENCHAUDHURI, P. and METHA, C. (2004). Exact analysis of dose response for multiple correlated binary outcomes. *Biometrics* **87**, 241–247.

HIRIJI, K. F. (2006). *Exact Analysis of Discrete Data*. London: Chapman and Hall/CRC.

HOCHBERG, Y. and TAMHANE, A. (1987). *Multiple Comparison Procedures*. Wiley, New York.

HOEFFDING, W. (1952). The large-sample power of tests based on permutations of observations. *Annals of Mathematical Statistics* **23**, 169–192.

HOGG, R. V. (1962). Iterated tests of the equality of several distributions. *Journal of the American Statistical Association* **57**, 579–585.

HOGG, R. V. (1974). Adaptive robust procedures: a partial review and some suggestions for future applications and theory. *Journal of the American Statistical Association* **69**, 909–927.

HOLM, S. (1979). A simple sequentially rejective multiple test procedure. *Scandinavian Journal of Statistics* **6**, 65–70.

HSU, J. C. (1996). *Multiple Comparisons: Theory and Methods.* Chapman and Hall, London.

HUANG, Y., XU, H., CALIAN, V. and HSU, J. C. (2006). To permute or not to permute. *Bioinformatics* **22**, 2244–2248.

ITO, K. (1969). *On the Effect of Heteroscedasticity and Non-normality upon Some Multivariate Test Procedures*, vol. 2. New York, Academic Press.

JONCKHEERE, A. R. (1954). A distribution free k-sample test against ordered alternatives. *Biometrika* **41**, 133–145.

KENNEDY, P. E. and CADE, B. S. (1996). Randomization tests for multiple regression. *Communications in Statistics: Simulation and Computation* **25**, 923–936.

KIMELDORF, G., SAMPSON, A. R. and WHITAKER, L. R. (1992). Min and max scoring for two-sample ordinal data. *Journal of the American Statistical Association* **60**, 216–224.

KLINGENBERG, B., SOLARI, A., SALMASO, L. and PESARIN, F. (2008). Testing marginal homogeneity against stochastic order in multivariate ordinal-data. *Biometrics* , (DOI) 10.1111/j.1541–0420.2008.01067.x.

KUDO, A. (1963). A multivariate analogue of the one-sided test. *Biometrika* **50**, 403–418.

LAGAKOS, S. W., WESSEN, B. J. and ZELEN, M. (1986). An analysis of contaminated well water and health effects in Woburn, Massachusetts. *Journal of the American Statistical Association* **81**, 583–596.

LANCASTER, H. O. (1961). The combination of probabilities: an application to orthonormal functions. *Australian Journal of Statistics* **3**, 20–33.

LEHMACHER, W., WASSMER, G. and REITMEIR, P. (1991). Procedures for two-sample comparisons with multiple endpoints controlling the experimentwise error rate. *Biometrics* **47**, 551–521.

LEHMANN, E. L. (1955). Ordered families of distributions. *The Annals of Mathematical Statistics* **26**, 399–419.

LEHMANN, E. L. and ROMANO, J. (2005). *Testing Statistical Hypotheses.* Springer, 3rd ed.

LENTH, R. V. (1989). Quick and easy analysis of unreplicated factorials. *Technometrics* **31**, 469–473.

LOGAN, B. R. (2003). A cone order monotone test for the one-sided multivariate testing problem. *Statistics and Probability Letters* **63**, 315–323.

LOGAN, B. R. and TAMHANE, A. C. (2004). On O'brien's ols and gls tests for multiple endpoints. *In: Benjamini, Y., Bretz, F., and Sarkar, S. (Eds), Recent Developments in Multiple Comparison Procedures, IMS Lecture Notes, Monograph Series* **47**, 76–88.

LOUGHIN, T. M. and NOBLE, W. (1997). A permutation test for effects in an unreplicated factorial design. *Technometrics* **39**, 180–190.

LUCAS, L. A. and WRIGHT, F. T. (1991). Testing for and against a stochastic ordering between multivariate multinomial populations. *Journal of Multivariate Analysis* **38**, 167–186.

MACK, G. A. and WOLFE, A. D. (1981). K-sample rank tests for umbrella alternatives. *Journal of the American Statistical Association* **76**, 175–181.

MANLY, B. F. J. (1997). *Randomization, Bootstrap and Monte Carlo Methods in Biology*. Chapman and Hall, London, 2nd ed.

MANTEL, N. (1963). Chi-square tests with one degree of freedom: extensions of the Mantel-Haenzel procedure. *Journal of the American Statistical Association* **58**, 690–700.

MARCUS, R., PERITZ, E. and GABRIEL, K. R. (1976). On closed testing procedures with special reference to ordered analysis of variance. *Biometrika* **63**, 655–660.

MARITZ, J. S. (1995). *Distribution-Free Statistical Methods*. Chapman and Hall, London, 2nd ed.

MARSHALL, A. and OLKIN, I. (1979). *Inequalities: Theory of Majorization and Its Applications*. Academic Press, New York.

MILLER, R. G. (1981). *Simultaneous Statistical Inference*. Springer-Verlag, New York.

MOLENBERGHS, G. and RYAN, L. M. (1999). An exponential family model for clustered multivariate binary data. *Environmetrics* **10**, 279–300.

MONTGOMERY, D. C. (1991). *Design and Analysis of Experiments*. Wiley, New York, 3rd ed.

MONTGOMERY, D. C. (2001). *Design and Analysis of Experiments*. Wiley, New York, 5th ed.

MOSER, V. C. (1989). Screening approaches to neurotoxicity: a functional observational battery. *Journal of the American College of Toxicology* **8**, 85–93.

MOSER, V. C., CHEEK, B. M. and MACPHAIL, R. C. (1995). A multidisciplinary approach to toxicological screening. iii. neurobehavioral toxicity. *Journal of Toxicology and Environmental Health* **45**, 173–210.

MÜLLER, A. (2001). Stochastic ordering of multivariate normal distributions. *Annals of the Institute of Statistical Mathematics* **53**, 567–575.

O'BRIEN, P. C. (1984). Procedures for comparing samples with multiple endpoints. *Biometrics* **40**, 1079–1087.

PATEFIELD, W. M. (1981). An efficient method of generating random $r \times c$ tables with given row and column totals. *Applied Statistics* **30**, 91–105.

PATEFIELD, W. M. (1982). Exact tests for trend for ordered contingency tables. *Applied Statistics* **31**, 32–43.

PEDDADA, S. D., PRESCOTT, K. E. and CONOWAY, M. (2001). Tests for order restrictions in binary data. *Biometrics* **57**, 1219–1227.

PERLMAN, M. and CHAUDHURI, S. (2004a). The role of the reversals in order restricted inference. *Canadian Journal of Statistics* **32**, 193–198.

PERLMAN, M. D. (1969). One-sided testing problems in multivariate analysis. *Annals of Mathematical Statistics* **30**, 549–567.

PERLMAN, M. D. and CHAUDHURI, S. (2004b). The role of the reversals in order restricted inference. *Canadian Journal of Statistics* **32**, 193–198.

PESARIN, F. (2001). *Multivariate Permutation Tests with Applications in Biostatistics.* Wiley, Chichester.

PESARIN, F. and SALMASO, L. (2002). Exact permutation tests for unreplicated factorials. *Applied Stochastic Models in Business and Industry* **18**, 287–299.

PETRONDAS, D. and GABRIEL, K. (1983). Multiple comparisons by rerandomization tests. *Journal of the American Statistical Association* **78**, 949–957.

POCOCK, S. J., GELLER, N. L. and TSIATIS, A. A. (1987). The analysis of multiple endpoints in clinical trials. *Biometrics* **43**, 487–498.

PODGOR, M. J., GASTWIRTH, J. L. and METHA, C. R. (1996). Efficiency robust tests of independence in contingency tables with ordered classifications. *Statistics in Medicine* **15**, 2095–2105.

POLLARD, K. S. and VAN DER LAAN, M. J. (2004). Choice of a null distribution in resampling-based multiple testing. *Journal of Statistical Planning and Inference* **125**, 85–100.

RASCH, D. and GUIARD, V. (2004). The robustness of parametric statistical methods. *Applied Stochastic Models in Business and Industry* **46**, 175–208.

ROBERTSON, T., WRIGHT, F. T. and DYKSTRA, R. L. (1988). *Order Restricted Statistical Inference.* Wiley, New York.

ROMANO, J. P. (1990). On the behaviour of randomization tests without a group invariance assumption. *Journal of the American Statistical Association* **85**, 686–692.

ROMANO, J. P. and WOLF, M. (2005). Exact and approximate stepdown methods for multiple hypothesis testing. *Journal of the American Statistical Association* **100**, 94–108.

ROY, S. N. (1953). On a heuristic method of test construction and its use in multivariate analysis. *Annals of Mathematical Statistics* **24**, 220–238.

SALMASO, L. (2003). Synchronized permutation tests in factorial designs. *Communications in Statistics: Theory and Methods* **32**, 1419–1437.

SAMPSON, A. R. and SINGH, H. (2002). Min and max scorings for two sample partially ordered data. *Journal of Statistical Planning and Inference* **107**, 219–236.

SAMPSON, A. R. and WHITAKER, L. R. (1989). Estimations of multivariate distributions under stochastic ordering. *Journal of the American Statistical Association* **84**, 541–548.

SHAFFER, J. P. (1986). Modified sequentially rejective multiple test procedures. *Journal of the American Statistical Association* **81**, 826–831.

SHAKED, M. and SHANTHIKUMAR, J. G. (1994). *Stochastic Orders and Their Applications.* Academic Press, San Diego.

SILVAPULLE, J. S. and SEN, P. K. (2005). *Constrained Statistical Inference. Inequality, Order, and Shape Restictions.* Wiley, Hoboken.

SILVAPULLE, M. J. (1997). A curious example involving the likelihood ratio test against one-sided hypotheses. *American Statistician* **51**, 178–180.

SIM, S. and JOHNSON, R. A. (2004). New tests for multivariate ordered alternatives. *Communications in Statistics: Theory and Methods* **33**, 2027–2039.

SPRENT, P. (1998). *Data Driven Statistical Methods*. Chapman and Hall, London.

TAMHANE, A. C. and LOGAN, B. R. (2004). A superiority-equivalence approach to one-sided tests on multiple endpoints. *Biometrika* **91**, 715–727.

TANG, D. I., GNECCO, C. and GELLER, N. L. (1989). An approximate likelihood ratio test for a normal mean vector with nonnegative components with application to clinical trials. *Biometrika* **76**, 577–583.

TERPSTRA, T. J. (1952). The asymptotic normality and consistency of kendall's test against trend, when ties are present in one ranking. *Indagationes Mathematicae* **14**, 327–333.

TROENDLE, J. F. (2005). Multiple comparisons between two groups on multiple Bernoulli outcomes while accounting for covariates. *Statistics in Medicine* **24**, 3581–3591.

WANG, Y. and McDERMOTT, M. P. (1998). Conditional likelihood ratio test for a non-negative normal mean vector. *Journal of the American Statistical Association* **93**, 380–386.

WESTFALL, P. H. and KRISHEN, A. (2001). Optimally weighted, fixed sequence and gatekeeper multiple testing procedures. *Journal of Statistical Planning and Inference* **99**, 25–40.

WESTFALL, P. H., TOBIAS, R. D., ROM, D., WOLFINGER, R. and HOCHBERG, Y. (1999). *Multiple Comparisons and Multiple Tests Using the SAS System*. SAS Institute.

WESTFALL, P. H. and YOUNG, S. S. (1989). *p* value adjustments for multiple tests in multivariate binomial models. *Journal of the American Statistical Association* **84**, 780–786.

WESTFALL, P. H. and YOUNG, S. S. (1993). *Resampling-Based Multiple Testing*. Wiley, New York.

Index

adjustment
  Bonferroni, 26, 113, 152, 192
  Holm, 26, 63
  Scheffé, 150
ANOVA
  one-way, 7, 106
  two-way, 133
attainable α-value, 7, 139

cardinality
  CSP, 137
  CSP statistic, 137
  of the permutation sampe space, 12
  of the test statistic domain, 7, 18, 159
  synchronized permutation, 127
  USP, 138
  USP statistic (even n), 139
  USP statistic (odd n), 138
central limit theorem, 193
closed testing, 27, 75
combining function, 14
  direct, 15, 119, 120
  Fisher, 15, 16, 120, 126
  Liptak, 15, 16
  max$T$, 15
  Tippett, 15, 16, 119, 120, 126
comparison
  all-pairwise, 149
  many-to-one, 63
  multiple, 59, 149
  pairwise, 63, 127
  post-hoc, 108, 113
concordance monotonicity, 53

conditional
  inference, 3
  Monte Carlo method, 106
  power, 46
conditionality principle, 2
confidence interval
  individual, 152, 162
  simultaneous, 149
contingency table, 39

data
  observed, 3
  pooled, 2
    insufficiency of, 3
    sufficiency of, 2
dependence, 10, 12, 13
  positive, 41
  positive quadrant, 41
  structure, 10
distribution-free property, 3
dominating measure, 2, 3

effect
  interaction, 134, 141
  masking, 180, 186
  sparsity, 176
exchangeability, 3, 4, 29
  characterization of, 3
experimental error rate (EER), 173, 176, 191

familywise error, see FWE
free combination, 35
FWE, 28, 29, 34, 56, 59, 75, 78

gatekeeping procedures, 63
global
    p-value, 14
    null hypothesis, 12, 13, 16, 22
    test statistic, 12, 14

Hadamard matrix, 174, 198
    (in) Yate's order, 198

imbalance, 73
individual error rate (IER), 173, 175,
    191
inference
    conditional, 2
invariance, 3
    property, 3
isotonic regression, 47

likelihood, 4, 6
    principle, 2
    ratio, 2
    ratio ordering, 41
    ratio test, 53, 85
linear
    rank test, 47
    test statistic, 44

MANOVA, 168, 170
marginal inhomogeneity, 72
mid-p-value, 48, 74
Monte Carlo
    iterations, 17
    two-sample algorithm, 21
multiple
    binary endpoints, 70
    comparison, 32
    testing, 25, 30, 152

nonparametric
    bootstrap, 78
    combination, 15, 118
    family of distributions, 2
normal plot, 173

odds ratio, 40
orbit, *see* permutation sample space
order restriction, 57
    tree, 58

partial

null hypotheses, 24
null hypothesis, 12, 16
permutation
    constrained synchronized (CSP), 136,
        156
    distribution, 5
    non-null distribution, 4
    paired, 111
    pooled, 111, 127
    sample space, 3, 11
    structure, 141
    synchronized, 110, 128, 133, 154
    test, 8
    unconstrained synchronized (USP),
        136, 156
population, 2, 5, 10

randomization hypothesis, 29
reference set, 3

screening experiments, 173
sparseness, 73
statistic
    Anderson-Darling, 56
    Hotelling's T^2, 10
    Kolmogorov-Smirnov, 99
    Mann-Whitney, 121
    permutationally equivalent, 6, 54, 108
    Snedecor's F, 105
    Student's t, 5, 87
step-down
    multiplicity corrections, 75
stochastic
    effects, 106
    ordering, 41, 51, 106
    ordering (multivariate), 66
    ordering (univariate), 116
study
    dose-response, 57, 62
    double-blind, 42
    experimental, 1
    observational, 1
sufficiency
    principle, 2, 3

test
    F, 128
    adaptive, 49, 58
    bootstrap, 80

calibration of, 185, 191, 192, 204
consonant, 93
direct χ^2, 54
dissonant, 75, 93
distribution-free, 3
exact, 1
Fisher's exact, 34, 56, 67
Hotelling's T^2, 85
intersection-union, 94
invariant, 3
Jonckheere-Terpstra, 99
Kruskal-Wallis, 105, 107, 113, 128
Loughin and Noble, 176, 185
Mack and Wolfe, 120, 129
McNemar, 11
nonparametric, 3
properties
  consistency, 7, 8, 17, 143
  exactness, 7–9, 17
  unbiasedness, 7–9, 17, 143
  uncorrelatedness, 140
synchronized permutation, 135, 154, 168

two-sample, 18
union-intersection, 86, 94
testing
  marginal homogeneity, 70
  noninferiority, 93, 95
  procedure, 31
  superiority, 86, 93, 95
theorem
  factorization, *see* sufficient statistic
treatment
  levels, 1
Tukey's honest significant difference, 108, 113, 150
two-sample
  location problem, 5

umbrella
  alternative hypothesis, 115, 123
  half, 124
  permutation test, 119
unconditional power, 50
union-intersection principle, 55
unreplicated 2^K design, 174, 180